We hope you enjoy this book. Please return or renew it by the due date.

You can renew it at www.norfolk.gov.uk/libraries or by using our free library app.

Otherwise you can phone 0344 800 8020 - please have your library card and PIN ready.

You can sign up for email reminders too.

6/21

My Penguin Year

Living with the Emperors –
A Journey of Discovery

Lindsay McCrae

HODDER

First published in Great Britain in 2019 by Hodder & Stoughton
An Hachette UK company

This paperback edition published in 2020

1

Paperback ISBN 9781529325478
Hardback ISBN 9781529325454
eBook ISBN 9781529325485

Typeset in Sabon by Hewer Text UK Ltd, Edinburgh
Printed and bound in Great Britain by Clays Ltd, Elcograf S.p.A.

Hodder & Stoughton policy is to use papers that are natural, renewable
and recyclable products and made from wood grown in sustainable
forests. The logging and manufacturing processes are expected to
conform to the environmental regulations of the country of origin.

Hodder & Stoughton Ltd
Carmelite House
50 Victoria Embankment
London EC4Y 0DZ

www.hodder.co.uk

To Becky, Walter and Ernest

Contents

3 April 2017

Standing motionless at the ice edge, I scoured the recently frozen ocean that lay below me and listened. The sky was clear, the atmosphere completely silent. Looking towards the midday sun that hovered just above the horizon to the north, falling glitter twinkled as it filled the air. 'Diamond dust', suspended ice particles, drifted by, hanging in the ether. There wasn't a breath of wind to disturb them as they rose and fell, shimmering in the dawn-like light. I raised my binoculars to my eyes. Straight to the north, black shimmering specks appeared like a mirage just above the horizon. At first just a few, but as my eyes adjusted, tens, then hundreds rose as if they were appearing from the other side of the Earth, advancing over its curve. Were the emperors actually returning? Was I watching them march towards me in great long lines like I'd imagined? If so, this was the day I thought I'd never experience. One of the most extraordinary sights Earth's natural world has to offer. This was the day I'd dreamt about as a child.

At −25 degrees Celsius, my face was numb and my cheeks were as red as my polar suit. My fur hat, wrapped around my face, had developed a crust of ice where my exhaled breath had solidified onto its fibres. But despite the discomfort of the conditions, the beauty of the landscape distracted my mind from any pain.

Grabbing the camera, I looked through the viewfinder. The enormous lens magnified the landscape further and compressed the distance between me and the horizon. As far as I could see, black specks swayed from side to side, seeming to waddle ever closer over the ice. This was it; the emperors were finally on their way. As they approached, the shape of each individual became distinguishable: rounded bodies, elongated necks and long, sharp beaks. Single-file lines of white bellies reflected the sky's light, illuminating the white snowy surface in front of them. Jagged icebergs, recently frozen in place, towered high, dwarfing the penguins as they passed under them. As I watched the penguins approach, the cold air rippled and the pink sky gave an ethereal feel. It didn't look real.

Long, winding parades crept their way towards me like an invasion of ants. At the front of each line, lead birds period-ically paused, causing a comical pile-up effect down the line. Extending their necks, each penguin looked impatiently over the shoulder of the bird in front. The recent snowfall had completely covered the ridges in the newly formed sea ice, making the surface look flat and uniform, and at speed, the penguins marched closer, leaving long, narrow tracks behind them.

The spectacle playing out in front of me was exquisite. If I had tried to describe the emperors' return, my portrayal wouldn't have come close to what I was witnessing. I felt like I was on a different planet and had no idea that this level of beauty existed.

I.

To *the* Other End *of the* Earth

My passion for the natural world started very young; in fact, I don't think I can remember a time in my life when I didn't love everything about nature. It sometimes feels like every memory I have of my younger years is linked to the outdoors; I was essentially outside every moment I could be. Growing up in Cumbria on the edge of the Lake District National Park was an absolute privilege. I lived next to an estuary and with high mountains just a stone's throw away, the range of habitats I could access was fantastic. It was idyllic. Like all children, I was born with a natural curiosity for the world around me and it just grew and grew. By the age of eight I had already decided I wanted a career filming wildlife. That might sound crazy but I just knew that was what I wanted to do. I spent all my spare time exploring my local woodland and riverbank and I had it practically all to myself.

Every day, I found something new, but my most fascinating find was a family of badgers in a little patch of woodland close to my family home. They captivated me from the moment I saw my first head poke above ground at dusk and although I didn't know it at the time, their private lives and shy-looking black and white faces were to be my ticket into the television world. I think I was a slightly unusual twelve-year-old, but it was what I adored doing, sitting quietly watching the natural world go about its business around me. I spent nearly every evening with the badgers getting to know their habits and felt incredibly privileged to have such intimate views of them. Each spring when the newly born cubs ventured above ground for the first time and played in amongst the carpet of bluebells, I felt at my happiest. Being in their company was addictive.

My obsession with wildlife led me to write a letter to the BBC's British wildlife programme *Springwatch* when I was just fourteen years old. *Springwatch* is a fantastic series that annually showcases the best of British wildlife and back when it started it was totally unique, capturing never-seen-before images of the wildlife I loved. Watching the programme was always a highlight for me and was the only thing that got me inside on a spring evening. Every year they based themselves in a specific location in the UK; obviously, thinking I had the most incredible local patch on my doorstep, I wrote to them detailing how perfect the Lake District would be for the following year's series.

It was basically my back garden, home to badger setts, fox dens, birds' nests and a large pond. I included a map with my letter detailing the exact locations of all the wildlife I had discovered. Not expecting a reply, I was surprised when I returned from school one day to a message on the answerphone. Impressed by my passion for badgers, they wanted to make a little film with me that would appear on the programme. I literally couldn't believe it! Of course I jumped at the chance and everything was arranged. The process of making a film, albeit short, fascinated me and even though I had no previous experience with a camera or any technical knowledge whatsoever, it gave me an understanding of how making television worked.

The *Springwatch* team lent me a camera and I spent every evening up in the wood with the badgers recording as much footage as I possibly could. It didn't take me long to realise that I could turn all my intimate wildlife encounters into unique filming opportunities. I found myself naturally filming all sorts around my badgers' sett: birds singing in the trees, bluebells blowing in the wind, even a snail working its way up a fallen log. They were the perfect ingredients for creating a film sequence on the life in the wood around the badgers' sett, rather than just a one-off shot of the badgers themselves.

With a combination of the footage I recorded and material from a film crew and producer who came to film me, *Springwatch* created a short film about a typical evening of mine watching the animals that fascinated me most. Head to toe in camouflage with an old army scrim net draped over me, I was shown with three badger cubs and their mother playing within just a few yards. Despite a bit of good-humoured stick from school mates about my camouflage appearance, I felt honoured that my passion had been recognised by such a well-loved television series and, being relatively young, I received countless messages from viewers saying how nice it was to see someone of my age on the programme.

Featuring on *Springwatch* just spurred me on more and I started saving every penny I could to buy a camera of my own. I'd got the bug and, with a short film under my belt, began to believe a career in the wildlife television world really *was* possible. It was my dream and I was prepared to do anything to achieve it. Thankfully, my mum took pity and gave me a loan so I could buy my much-desired camera. I bought it second-hand and can remember the moment it was delivered. It was mine and now I could finally record all the amazing things I was seeing out and about. Spending every second of my spare time outside filming as much wildlife as possible, I started to develop some much-needed technical skills.

'It's now sixty degrees *below* zero. Abandoned by the sun, the males are left alone with their eggs to face the coldest, darkest winter on Earth.' I can literally pinpoint the moment my dream became fixed in my mind. It was 5 November 2006, and I, along with millions of others, was glued to the television watching David Attenborough's groundbreaking wildlife series, *Planet Earth*. I'd waited weeks to see the series and I wasn't going to miss it for the world. Emperor penguins breeding on the Antarctic ice sheet filled my television screen, battling the atrocious conditions that a winter at the bottom of our planet threw at them. Goosebumps rippled from my neck down my back and, seeing the penguins' faces covered in ice, I shivered.

The ten-minute *Planet Earth Diaries* at the end of the episode focussed on how the emperors had been filmed. I felt I was there when in reality I couldn't have been further away. I was sitting on the sofa with a cup of hot, steaming tea in front of me. I felt an urge to be there, to be in that position, experiencing all those sensations. Even though I was young, I knew I wanted a career filming wildlife, although nothing was certain. It still felt like a fantasy, however, and never in a million years could I envisage myself achieving it.

In my last few years at school I spent every spare moment filming. Before school; after school; I even snuck out during the lunch break on occasions. From making my *Springwatch* film, various contacts I'd made at the BBC kindly studied tape after tape of footage, giving me advice on where I could improve. Leaving school at eighteen, I'd lost all my academic motivation and all I wanted to do was be out in the fresh air watching the natural world go about its business. Incredibly, during my first year out of education, my name had been mentioned so frequently within the BBC *Springwatch* office that I received a call offering me a job. I'd made a conscious decision not to go to university, but with nothing lined up within the industry that I desperately wanted to enter, it couldn't have been more perfect.

After a meeting over a cup of tea in Bristol, *Springwatch* gave me a job as a 'runner', basically an odd-job man who was there to do anything asked of me. Spending over a month making teas and coffees and visiting supermarkets three times a day was strangely good fun and it got me a foot in the door, mixing with the right people and showing them how capable and keen I was. My stint on *Springwatch* led to camera-assisting jobs and opportunities to get hands-on experience with professional gear. It was just what I needed to break into the industry and from then on I was employed on various short-term projects that gave me further experience.

I couldn't believe the opportunities that were coming my way. I felt so unbelievably privileged and, after a few years, my career had well and truly kicked off. I had travelled all over the UK and had started spending more time filming abroad, from wolves in Alaska close to the Arctic Circle, to armadillos in the Brazilian Pantanal almost on the equator. I was living my dream. Working freelance meant I didn't know where I'd be sent to next, which was exciting as it could have been literally anywhere on the planet to film absolutely anything. It was always a surprise.

Throughout this time, I held on tight to my dream of going to Antarctica, but filming opportunities there were few and far between and so it still eluded me. It was one of only two continents I'd not managed to reach and I was wondering if I'd ever get there.

'Big BBC film trip.' Opening up my emails, the subject line caught my eye immediately. The email was from Miles, a producer at the world-famous BBC Natural History Unit. I had never actually met Miles so it was a surprise to get an email from him, but I knew exactly who he was. Growing up, natural history television producers were to me what Premier League footballers were to my friends. Something about this email felt different. Normally, I would just get a call or email asking if I wanted to go somewhere on a certain date, but this was strangely vague. It was asking for a call, but didn't give me any more details. My interest was piqued immediately. Having arranged a conversation over the phone one afternoon, Miles still hadn't given me any clues so there was a certain amount of excitement when I answered his call. We were having a small extension built onto the back of our house and I had spent the day carting heavy lumps of stone and old kitchen units down the drive. Covered in dust, I stood at the top of my garden for mobile reception (not easy to locate in the Lakes) to take Miles's call.

'Lindsay, glad I've caught you. I've not got long but just quickly, how would you fancy emperors in Antarctica?'

Falling silent, I was stunned. I was being offered my dream job! Was this real? Already picturing myself covered in ice in a raging Antarctic blizzard, I was just about to say yes, of course I fancied it. Miles quickly butted in . . .

'There's a catch, mind you: it's an eleven-month trip. We would send you down December '16 and bring you back November '17.'

I didn't know what to say.

Again, Miles quickly chipped in, 'Don't give me an answer now, go and think about it. I can give you two weeks.'

'That sounded serious!' Jonny our builder shouted across the lawn. I kept my mouth shut. I didn't think I should mention it to the builder before telling family.

From the minute I put the phone down, my mind raced. All I could think was that I wanted to go, but life is never that simple. I had recently bought a house with my girlfriend, Becky, whom I had been with for six years. She had uprooted her whole life to move up to the Lakes to be with me and we were living so happily together with our two dogs, Willow and Ivy. Becky used to work in television so she knew the deal with me being away, but what I was about to ask was off the scale. We were used to being apart for a month or five weeks at a time, but almost a year was unheard of. It was very much a decision I needed to think long and hard about. Was this opportunity worth the sacrifice of being away for that long? What would Becky and my family think? At the same time as feeling huge excitement, I felt physically sick. This was my ultimate dream and I was being given the option to make it real, but at what cost? Things like this just don't happen to people like me. I'd been given two weeks to think; the clock immediately began to tick. I felt as if I had under a fortnight to decide my entire future. I knew it would be the most challenging thing I'd ever experience, but that's what excited me. It was extreme in every way: the furthest I could travel from home, the longest duration away from home and the most challenging conditions I'd ever work in. A career-defining project versus risking losing everything in my personal life. What a decision to make.

I began to do my own research. The deeper I probed, the more I felt I couldn't give up this opportunity. Although penguins had always fascinated me, I'd never actually seen one, not even in a zoo, but like the majority of people, they held a special place in my heart. There is just something about penguins that I can't quite put my finger on. If penguins as a creature were something special, then emperors as a species were on a whole other level. In nature, nothing quite defines perfection like an emperor penguin. Out of the world's seventeen species of penguin, they're the largest at up to 115 centimetres, and they're the heaviest, tipping the scales at almost twenty-five kilograms.

Emperors are the most recognisable of all penguins. Immaculately dressed as they are with their golden cravat, a dazzlingly white diamond belly and a long pink-and-blue curved bill, in my opinion they're one of the most beautiful animals that exist on Earth. With forty-four confirmed colonies of emperor penguins breeding around the coastline of Antarctica, it is estimated the population of emperors sits around the 600,000 mark. In 2009, an emperor penguin census was undertaken from space using the advance in technology offered by satellite imagery. Before that, a count from the ground in 1992 had revealed just half that figure. It wasn't just the number of individuals these new high-resolution photographs revealed, it was previously undiscovered and unexplored colonies that came as the surprise. Dark guano stains that showed up against the bright white snow and ice were a clear indication of an emperor colony being present. My head was filled with icebergs, adventure and penguins. I suddenly had a one-track mind. I was going.

Becky had been aware of the email from Miles as soon as it had come through and was eager to know what the job entailed. She always supported me and got excited about any opportunity I was offered. Knowing I had a call booked in with Miles that afternoon, she went off to work. After my phone call with Miles I realised I didn't have long to make this life-changing decision, so had to think fast. I needed to broach the inevitably touchy subject with Becky as soon as possible. I thought over dinner would be best but deep down I knew her reaction wasn't going to be great; it was a real shock to me and it was going to be a shock to her too.

We had been apart for a big part of the year already and it had definitely been difficult at times. She was always the one getting left behind, the one holding everything together at home so I could concentrate on work and live my dream. On one occasion she had made me get my diary out so we could see just how much we were apart. I think it shocked us both that we had spent over six months of the year in different locations. When I was home, life couldn't have been better. Was I really just about to turn all this upside down? I was well aware that she'd been the one to make sacrifices by moving from the Midlands to the Lake District where I lived, rather than the other way round.

Despite having only a microwave in the garage and a small metal barbecue at my disposal while our extension was being built, I'd produced a feast. The evening was beautiful and my recently completed patio at the top of the garden looked as if it was ready for its christening as the last ray of sunshine illuminated its golden slabs. I wasn't normally romantic, but foolishly I'd gone all out, thinking it would soften the blow. I'd wedged an old cane fishing rod into the drystone wall and hung a small lamp from its tip. Its weight produced a bend in the rod more serious than a fish ever had and it hovered just above head height in the centre of the table. Occasionally, it swayed from side to side in the summer breeze as two candles flickered in a couple of old jam jars underneath.

Becky returned from work, the clatter of the side gate alerting both dogs, who'd been helping me set up. Bounding down the garden in excitement, they revealed where I was. She took one look and knew I was up to something. As the sun began to dip below the hills, we sat to eat.

'What is it?' she asked as a subtle smile appeared, giving me hope. A moment of silence; her eyes stared right through me. 'Go on. How long?' she continued and without giving me time to respond she began guessing. 'Three months? Four? Five?'

She got to six and the smile had vanished. A look of horror replaced it, then a gulp and the awkwardness of shock.

'Emperors, Antarctica,' I replied, thinking this might make her understand.

'How long?!' The species didn't appear to matter as she responded in a stern tone.

'Eleven,' I replied.

Pushing her chair backwards, she discarded her scrunched-up serviette I'd so neatly folded just half an hour before. 'Absolutely not! I can't believe you've even contemplated asking me!' She disappeared down the garden towards the house with Willow and Ivy tight against her heels. I guessed that was a no then.

With just a couple of weeks to make a decision, I felt under pressure. I began to seek opinions from other people I knew who could offer me advice. I spoke to a television producer, a close friend and someone who'd been to Antarctica and done their own time living down there. Various troubling stories of people returning from isolation kept cropping up, but it was clear the challenge my relationship would face would be one of the toughest problems.

I drove to Bristol and met with Miles to get more information about the project. The trip was by no means confirmed but they needed assurance from me that I'd be willing to commit so they could start further planning. Unbelievably, the conversation with Miles did the opposite of what I had expected. The more I heard, the more I began to think that maybe this trip wasn't actually for me after all. The plan was to live at the German research station Neumayer III. Accompanying me would be just two other team members, the three of us living alongside nine German scientists. Miles warned me that temperatures could reach −50 degrees Celsius and for eight of the eleven months I'd be completely isolated. No way in; no way out. The weather becomes so unpredictable and extreme that both air and boat transport would be unable to reach me. The penguins were beginning to sound like the only part of the project that appealed to me. I got the impression Miles could tell I wasn't as keen as when we had first spoken on the phone. I was used to working alongside people I'd never met, but this was different.

'Would I know anyone else on the team?' I asked. With the project still in its early days, Miles was slightly reluctant to answer, but he eventually gave in.

'Will Lawson. Do you know him?'

Will and I had worked together a few times making wildlife films in the UK. He was a funny guy and well liked. Learning Will was on the team made me feel a little more optimistic, but I still drove home not knowing what to think; on the one hand it sounded too extreme both mentally and physically, but on the other hand I knew this was the most incredible opportunity I'd ever get. I wasn't sure I could let it slip away. How many people get asked to spend a year living alongside the most charismatic and beautiful animal on the planet and then share it with millions of others on screen? I arranged more calls and further researched living through a winter of isolation. I wanted to know more about the mental side of things and whether I could convince myself I could actually endure it. The psychological aspect that Becky would experience if we were to be separated for such a prolonged period still hadn't dawned on me. But selfishly, I still couldn't see past those iconic *Planet Earth* images. That could be me. I was nervous, but I had to go for it. I just needed to convince Becky.

With only one evening remaining before I'd promised to give Miles an answer, Becky returned from work. Slow streams of raindrops crept down the lounge window; no sunlit dreamy banquet that evening. I'd been working on a speech about what I'd say to her all day, hoping to bring up the topic again. We hadn't spoken about it since I first broached the subject but I felt the tension starting to bubble up in the background. On an A4 piece of paper folded into quarters stuffed into my trouser pocket, I'd bullet-pointed the benefits and rewards we'd both get if I went. As I sat down on the sofa, Willow and Ivy could sense my nervousness. They jumped up and settled together between us. The television was turned off and all I could hear was the rain getting heavier.

Before I could speak, Becky announced, 'You can go! We'll make it work. Don't worry.'

'Really?' I replied in shock.

'Yes. Now what's for tea?' she said, smiling, wanting to change the subject as quickly as possible.

My first proper encounter with Becky had been in 2010 in Richmond Park in central London filming for *Springwatch*. We had actually crossed paths five months before in a production office in Bristol; however, she hadn't remembered that! In the bar following the first day of filming, we'd got chatting. Typical television talk ensued: favourite places, previous adventures, dream assignments. I explained in detail my dream to her: filming emperor penguins in Antarctica. Becky had never forgotten. The fact she had said yes says a lot about who she is. I think people told her she was mad, but our families and friends supported us and that was all that mattered.

The following morning I rang Miles to tell him I was up for

it and I spoke to Will for the first time about the project. Knowing Will already was a huge factor in my decision-making process. Although I loved meeting and working with new people, I'd realised I needed reassurance that I wouldn't lose my sanity during my eleven months away. I felt comfortable around Will and knew we'd be able to reassure and encourage each other during the tough times. I also knew we'd have great fun. No matter how strong I felt I could be, tough times were inevitable. His decision to go couldn't have been a straightforward one either. We spoke about what we were most excited about, the unique nature of our task and what kind of film we could return with. Once I had committed, Miles immediately got me more involved, putting me in touch with the third and final member of our crew.

Stefan, a German photographer, had actually travelled to Neumayer before as a scientist. Working as a geophysicist, in 2012, Stefan had lived in Antarctica for fourteen months and had become fascinated by the penguins that lived next door. It was good to have someone to talk to who knew what to expect and on a large notepad I started filling pages with questions to put to him. The hugely ambitious project was still by no means confirmed but having the commitment of a core filming crew, the production team were free to start making further plans such as applying for filming permits and discussing with Neumayer whether they could realistically provide the three of us with the support we required. The team's intentions were clear and, despite the whole idea seeming crazy, everyone involved wanted to make it happen.

With things coming together on the filming side it was time to focus on my personal life. I was keen to make my intentions with Becky very clear. I may have been going away for a year but I had no intention of losing her along the way. Having been with her for six years by then, I wanted her to know how much she really meant to me. Although she'd told me she was willing to wait, I needed to make sure she knew it was worth it.

Just before Christmas, almost one year before departure, I proposed. Asking her to marry me felt fantastic but I had no idea whether she intended getting married before I left or after. We only had one year before I would be gone, after all. I think I was a bit naïve. Becky had absolutely no intention of waiting until I was back and began planning immediately. I think maybe it was the perfect distraction from the upcoming trip.

I, on the other hand, was busy researching both emperor penguins and Antarctica; not particularly helpful for wedding prep, but it kept me busy and out of the way. Antarctica is Earth's southernmost continent and one of our planet's last great natural wildernesses. Ninety-eight per cent of its surface is hidden, locked under a frozen layer of ice comprising, unbelievably, approximately one-fifth of the world's fresh water. It's the coldest and windiest place on Earth and, covering an area almost twice the size of Australia, it's the fifth-largest continent.

Surprisingly, it's also the driest, experiencing only one to two inches of precipitation each year. Antarctica holds the record for the coldest temperature ever recorded on Earth at -89.2 degrees Celsius and even throughout the summer months the temperature rarely rises above zero. Despite it being one of the most hostile places on Earth, humans do have a presence there, mainly in the form of scientific research stations dotted around the edge. As well as these coastal bases a few stations exist at altitude and inland, including the famous Amundsen–Scott base at the South Pole. In total there are forty-four stations that are occupied throughout the year, with a further thirty smaller stations manned only during the summer months.

There were still twelve months to go until we actually left for Antarctica but there was an incredible amount to do. I hadn't quite appreciated the amount of work that had to go into organising a trip like this. It was a massive risk to send us down there, with no guarantee of what we would get on film, and the team were keen to mitigate as many risks as possible before we left. To tell the emperors' story in a more unique way both visually and behaviourally, the German research station Neumayer III, run by the Alfred Wegener Institute for Polar and Marine Research (AWI), had been chosen as our base. Sitting on an ice shelf rather than solid ground and having one of the largest colonies of emperors on Antarctica, numbering around 10,000 individuals within ten kilometres of the station, it was the perfect location.

Previously filmed colonies had been surrounded by enormous cliffs of ice and rock and the fact that the colony at Neumayer didn't have any of these would straight away make the film look different. Not only was it an incredible setting, but the facilities that Neumayer could provide to accommodate us throughout the year seemed perfect. I hadn't really thought about anything beyond saying yes to the trip. Having also never worked or experienced the hostile conditions that I was going to have to endure, I had a lot to learn. An enormous amount of training would be required to keep me safe and the next twelve months were going to be extremely busy.

Being confined to such a remote location where rescue was

unlikely came with huge risks. Neumayer had one of the most advanced medical facilities in Antarctica, but I still had to prove I was physically fit enough to last the eleven-month trip including the eight months of isolation. Until relatively recently, it wasn't uncommon for people travelling to Antarctica to undergo surgery to have their appendix removed before travelling, to avoid any possibility of needing an appendectomy while away from medical help. Nowadays, facilities are so advanced at certain stations that those operations can be performed in Antarctica. Regardless, I was examined from top to bottom, literally. Urine samples, stool samples, X-rays, ultrasounds and blood tests.

I started to learn things I didn't know. One scan revealed that my two kidneys were connected, a common condition known as a horseshoe kidney. 'If you've never had one before, you can find funny things when you have scans,' said the sonographer, seeing my nervous-looking face. I had a thorough dental check, resulting in a filling being replaced to avoid the possible event of it coming loose. To my horror, an X-ray of my mouth almost resulted in having my wisdom teeth removed *just in case* they became impacted while I was there. Luckily, I got away with that one!

There followed an exercise electrocardiogram (ECG) in Liverpool and an eye test at my local optician's. The amount of information needed to assess whether I was strong and healthy enough to travel was exceptional and the enormity of what I'd entered into started to become apparent. Meetings at the AWI headquarters in Bremen in northern Germany began appearing in my diary, including environmental conferences and a polar clothing fitting for all my specialist garments. Due to working outside in temperatures that had the potential to drop to as low as −50 degrees Celsius, the AWI provided me with an enormous collection of gear. I was fitted with so much that I probably could have got away with not packing any items of my own! Hats, gloves, mittens, thermals, boots, polar suits, jackets, socks, goggles, glasses, lip balm, sunscreen. Everything had been thought of and I was incredibly well provisioned. I attended an in-depth firefighter training course in the unlikely event Neumayer caught fire, and a two-day course in the UK run by a member of an air ambulance crew gained me imperative wilderness medical skills to ensure that, should I have to administer first aid to anybody in the field, I was capable of doing so. It was intense, but having information thrown at me from all directions ensured that during my stay in Antarctica I'd be prepared for any scenario.

It felt strange, but we had barely talked about what we were actually going to film. The penguins hadn't been mentioned for months. With the amount of medical attention I was receiving, I began to wonder what would happen if anything *did* go wrong, especially in the middle of the winter during the period of lockdown. I looked back though the documentation the AWI had provided me with.

As part of the core team that stays during the winter, there would be a doctor living with us who would have surgical qualifications and experience. The medical facilities on the station were state-of-the-art and would offer the ability to treat most scenarios. Whether it be just a cut finger or a more serious injury, they were prepared. The equipment was so advanced that they had a telemedicine system in place: in the event of a serious incident, live patient data could be displayed to medical specialists back in Bremerhaven hospital in Germany while the patient lay on a bed in the Neumayer surgical room at the other end of the planet. It reassured and terrified me in equal measure. It was serious stuff. So serious that I didn't dare mention it to Becky; I thought some things were best kept to myself.

It was made clear, however, that not everything could be done at Neumayer. Medical evacuation was a last resort, but in a place with no planes, the likelihood of actually getting anyone out was almost nil. Unbelievably, I discovered that it would be quicker and easier to evacuate someone from the International Space Station than from Antarctica! It was a risk I just had to live with.

During the height of summer I travelled to Austria for a week-long mountain-rescue training programme. At over 3,000 metres in altitude, Taschachferner was a remote glacier that retained snow and ice all year round. The evenings were spent learning how to use global positioning systems (GPS) effectively, how the day-to-day running of Neumayer III station worked, details about the facilities and some house rules. During each day we climbed for an hour to reach the ice. Rope work in deep crevasses was an amazing experience and I was learning new skills in conditions that were as close as we could get to those in Antarctica without having to travel there. Acquiring techniques that enabled me to recover other people as well as myself in emergencies was great fun and taking part with the eleven other members of our overwintering team was an important bonding exercise. We really were starting to feel like a team.

I began to work out who I could play jokes on when we were living together and who wasn't quite ready for my slightly immature humour. It immediately became clear that we had a great bunch of people who were up for a laugh and they were all equally interested in my job as I was in theirs. On the penultimate day of the training course, I was lying next to Will soaking up some sun during a break in between exercises. All of a sudden I received a text message on my phone. We were so remote and at altitude, I was surprised I had any reception. Opening the message from Becky, I saw it read, 'Can you talk?' Worried, I quickly rose to my feet. Had something happened? Was she having second thoughts about me going away? I walked away from the group to give her a call. We hadn't managed to speak for a couple of days, which heightened my feelings of anxiety. Within half a ring she answered.

'Are you OK?' I asked.

With not another word, she replied, 'I'm pregnant.'

I couldn't believe it. It was the most incredible feeling in the world, knowing I would become a father. I felt ecstatic, nervous and terrified, all at the same time. It put a whole new perspective on my trip south.

Becky and I had always wanted children and we worried about delaying starting a family even longer because of the trip. It was a case of either waiting nearly two years for me to return before trying for children, or trying before and seeing what happened. Having our first child while at different ends of the planet was something we'd thought about a great deal but I never expected it to become a reality. I had already committed; there was no option other than to be apart for the birth and the baby's first seven months. It was a scary prospect: Becky would have to single-handedly bring our child into the world and for seven months look after it alone, and I would have the mental and emotional challenge of being separated and missing those first moments of my first child's life. I really was racking up the sacrifices we were making for this trip, but I had no other option.

I kept the news to myself, not telling anyone else involved in the trip. I felt that if Miles heard, he'd pull me off the project. He'd have had every right to. The mental strength needed for an eleven-month trip with eight months of isolation was enough to deal with, let alone missing the birth of my first child. The decision was mine and Becky's and knowing we'd make it work was all that mattered. I didn't want to worry Miles or anyone else involved with something that was out of their control; they had enough to think about.

After almost two years of planning, a departure date was confirmed: 16 December 2016. Becky and I had made the most of our last few months together. Following our unforgettable honeymoon, we took off abroad to catch some autumn sun. I'd never been keen on sunshine holidays but as it was my last chance to feel the lovely warmth of the sun for a while, I thoroughly enjoyed it.

Once I got home, however, the reality of what I'd signed up for properly sank in. I started to have second thoughts. For weeks, I woke up in the middle of the night in cold sweats. I'd find myself staring at Becky asleep next to me wondering if I could really leave her on her own for that long. It made me feel sick. Could I actually cope with the mental aspect of being abandoned so far from home? I wasn't convinced I could, and started seriously thinking of how I could get out of it. What on earth was I doing? Despite having committed to the trip over a year before, I just wanted the whole thing over and I came seriously close to giving in.

I'd been incredibly well examined physically but nothing regarding my mental state was ever mentioned. Surely demonstrating that psychologically I was strong enough was an imperative part of the process? Not just for my benefit but also for the group of people I'd be living alongside. I contemplated asking Miles whether I should take some formal tests, but I kept quiet in case I didn't pass them. Some days I believed I was strong enough, others I didn't. The fact that I was having second thoughts made me worry, but I kept it to myself. I didn't want anyone, especially Becky, to find out I was struggling. The trip hadn't even started and the ups and downs had already begun.

My final week at home came around too fast. We'd had to make some pretty big decisions over the autumn and with my departure looming, Becky had reluctantly left her job, enabling us to spend the last few months together. She also made the decision to move back to Northampton to be with her parents while I was gone. Giving birth and bringing up a baby on her own in Cumbria simply wasn't an option without her close family nearby. Out for one last walk with the dogs, I realised it wasn't just *my* last moment of Lakeland freedom. Willow and Ivy would travel south with Becky and even though they'd be well looked after, the landscape in the Midlands just wasn't the same. Watching them running across the fell side, I couldn't help but feel extremely guilty.

All packed and ready, on 15 December I said goodbye to my parents. It proved a lot tougher than I'd expected. I'd only ever seen my dad cry once before. He was mentally pretty tough but as he lived only a mile up the road, I'd seen him almost every day when I was at home. If not meeting for a coffee, we'd pass each other on the road and wave. My mum had recently moved house from our original family home to just a few hundred yards away and there weren't many days I hadn't seen her in the village either. The eleven months in the Antarctic would be the longest I'd have gone without being at home, but it was more the prospect of isolation that put a whole new perspective on the trip.

Becky and I drove south down the motorway to spend our last evening together at her parents'. My sister and her partner, not living too far away, popped over for dinner, giving me the chance to wish them well. My sister and I had hated each other as children but since our late teens we had become extremely close, so much so that she was my 'best woman' on my wedding day.

The following morning we woke to one of the hardest days of our lives. Having packed my car I hugged Becky harder than I ever had before, squeezing her so tightly that she told me to ease off. I bent down to kiss the dogs. 'Willow, Ivy. Look after your mum,' I told them. Ivy glared at me through the tops of her eyes. They both knew I was leaving. I waved at my sobbing wife through the window, a heavy-hearted image that would never leave me.

Driving to Heathrow in floods of tears, the only thing that prevented me from turning round was reminding myself how fortunate I was. I was at work now and had to put my emotions aside. Will, who'd made his own way to the airport, greeted me with a tight hug. 'Let's go and get this job done,' he said supportively.

Our journey to Antarctica was by plane. Out of the three main routes, the quickest way to Neumayer was via Cape Town in South Africa. Flights from Cape Town operated to the Russian airbase Novolazarevskaya (or 'Novo'), an ice runway directly south from Cape Point.

We arrived in Cape Town with time to spare as the unpredictable weather in the south meant flight times could be shifted, sometimes by days. For a total of four days I walked aimlessly around the busy harbour. The Antarctic weather had apparently been a little too rough for planes to land and it had prevented our flight from leaving on time. The enormity of leaving Becky for a year was still raw and it was all I had thought about since leaving home. Will and I had met up with Stefan and our nine fellow overwinterers, but feeling so low, I found it difficult to bond. They were all extremely excited and made the most of being in such an iconic city, but I found it very difficult to think about anything other than Becky and our baby, and knowing that there would be no turning back as soon as I stepped onto the plane to Antarctica, I just wanted to get on with it before I had a chance to change my mind.

The only thing that allowed my mind to drift was glimpses of swallows that swooped around buildings and fed on flying insects. As they were migratory and spending the summer in the UK, I knew they'd made the same journey south that I had. Back home, I loved seeing the first swallow arrive back in the spring and knowing I wouldn't see another for well over a year, it was lovely to have the chance to say goodbye and wish them a warm and lucky summer without me.

Finally, on the longest day of the year (south of the Equator), I found myself looking up at the departures board in Cape

Town International Airport. 'VDA 9018 – Antarctica – Gate B1-2.' Still feeling slightly sick at the prospect ahead of me, I boarded the bus that would take our team across the runway to the plane. With four huge engines slung beneath two broad wings, a blunt glass nose and a horizontal tail, flight 9018 looked as if it was about to take me into space. Operating up to twenty flights between Cape Town and Antarctica each year, the IL-76 was an aircraft designed by the Soviet Union in the late 1960s to deliver heavy items to undeveloped areas. Its ability to land on unpaved runways and carry huge amounts of freight via a ramp access at the rear made it the perfect choice to be flying to such a remote location. As I boarded I took some photographs and, like an astronaut, took one last look behind me from the door. Strapping myself securely into what looked like an old ejector seat salvaged from the tip, I started to feel excited.

The interior was a hollow metal tube with a host of international flags hanging from its walls. A grey Portaloo was ratchet-strapped to the side of the aircraft behind the back row of seats, the website and contact number of its loaner in big white letters. I glanced across at a spare tyre and wondered what use that would be at 30,000 feet above the South Atlantic Ocean. I couldn't comprehend how big the jack would have to be!

Separating the cargo at the rear was a thin white rope net. With everything strapped down, I wouldn't have been surprised if we experienced weightlessness at some point! With only four small windows there were very few possibilities to see outside and as the doors were closed and the engines fired up it became obvious why my seat had a small bag of foam earplugs left on it. It thundered! I couldn't even hear my own voice. As the aircraft juddered into life, taxiing down to the end of the runway, a large television screen at the front flashed into life. A camera mounted on the aircraft's nose sent a live feed that offered the passengers their in-flight entertainment. It was the cockpit's eye view and it was perfect.

The engines' noise increased and everything vibrated. In what seemed like slow motion, we began our take-off. I could see the end of the runway on the screen and just when I thought we were going to run out of tarmac, the plane hopped into the air, front and rear wheels losing contact with the ground at the same time. The plane didn't just look weird, it felt it too, but I was on my way to the emperors and for the first time it felt good.

About halfway through the six-hour flight I left my seat in the second row and walked to the front where, in one of the emergency exit doors, there was a small circular window, no larger than a football. I leant forward to look down, the sun's glare forcing me to squint. Ice! Like smashed tiles, blocks of broken ice led my gaze into the distance. At 30,000 feet, it was my first glimpse of 66 degrees south, the Antarctic Circle. Up until this point every aspect of my preparation had brought my adventure closer to reality, but the sight of ice had huge significance. It finally started to feel real. I'd been questioning myself for months about whether I'd made the right decision to go, but all of a sudden, here was my answer. Mesmerised as the ocean below transformed from a deep-blue liquid to a blinding white sheet, the first wave of realisation hit me. I found myself staring down, attempting to focus my eyes on every single speck of ice. All I wanted to do was jump down and pretend they were stepping stones. With my earplugs still deep in position, I was in a world of my own and for the first time in a couple of months, I felt happy. I was excited and, with the feeling of relief that my trip was finally under way, relaxed.

As we approached ever closer, the screen at the front showed weather information at our destination: −2 degrees and clear skies. Tropical, I thought, for Antarctica but not quite the South African shorts and T-shirt weather I'd left behind. Grabbing my flight bag containing my cold weather clothing, I rushed back over to my seat to get changed. Opening my bag felt like Christmas Day as a child. Since the fitting of my polar clothing back in August in Germany, I'd waited for this day. Finally, the chance to put it to use. As I slid my legs into the all-in-one down suit and hoisted the straps over my shoulders, I felt like a proper explorer. Landing on the bottom of the planet was just moments away.

Three steps, two steps, I counted myself down the ladder as I disembarked the plane. Like Neil Armstrong on the surface of the moon, I jumped, the soles of my steel-toecapped flight boots landing hard on the ice. In front of me was a vast expanse of blinding whiteness and dark-blue sky and in the distance two grey rocky peaks protruded from the surface of the snow. The air was crisp and I could taste its purity. Like no other, the landscape stretched away in front of me; no trees, no brick buildings, no sheep. I turned to Will: 'We've made it, we've bloody made it!' We both laughed in excitement.

Passengers, the majority of whom were also first-timers, scattered themselves everywhere across the ice runway surrounding the aircraft taking photographs. Ground staff in green ski suits and orange high-vis jackets got to work servicing and unloading the aircraft. The temperature was incredibly mild and straight away I could feel my skin burning under the high midday sun. I felt daft in my polar suit and thick hat and within minutes I was overheating.

Standing on my own, watching, I thought about home and wondered what Becky would be doing. She was four months away from giving birth and I was already wondering what our little boy would make of his dad's trip. Even though he hadn't yet been born I recorded him a short video for when he was a bit older. With the plane in the background, I spun around to look in a northerly direction. I was keen to know in which direction home was, but all of a sudden I realised something. I rested my chin on my chest, looked down to my feet and waved at the ground. Home was, of course, at the opposite end of the planet.

Novo airbase and Neumayer III research station are positioned on the Queen Maud Land region of Antarctica, one of eight territories across the continent. The equivalent of a county or state back on the mainland, the 2.7-million-square-kilometre area was claimed by Norwegians in 1939 after they first set foot on it in 1930. The majority of the area is covered by the thick Antarctic ice sheet with a few mountainous areas of bare rock positioned inland protruding out of the snow. A settled weather system holding strong over the coast meant the final three-hour flight west to Neumayer could thankfully go ahead as scheduled. Having struggled with my first few days away from home, it felt as if it was meant to be. No hassle. Travelling around Antarctica in such consistently kind weather was rare, and the fact I was getting to Neumayer as quickly as possible gave me no opportunity to turn back. Along with the rest of the team I was hurried into a smaller aircraft, which had skis rather than wheels, waiting by the runway.

A large bearded Canadian pilot sat at the door to his aircraft and briefed us. I felt as if I was on a school trip. Our thirty camera cases followed on the trailer of a snowcat, but due to a miscommunication, there was no room for our half a tonne of extra baggage. 'We'll fly them to you tomorrow,' he said in his stern Canadian accent. Thousands of pounds' worth of high-tech camera equipment sat abandoned on the ice. I looked at Will, confused. I felt we couldn't leave without them. Reminding me of an old teacher, the pilot, Chuck, wasn't the friendliest of characters. From his demeanour, he had clearly flown his plane in Antarctica before and he didn't have much patience with a load of Antarctic newbies.

'Can we take just a couple, the camera at least?' I asked him, but no. Not a chance. I'd never met Chuck before but already I didn't like him. As frustrated as I was with how unhelpful he was being, we had no option but to trust him and leave them behind.

As the plane taxied back onto the ice runway, I watched as the trailer-load of camera cases sat neglected and alone. What if I didn't see them again? All the planning and preparation had gone so smoothly, I should have known some hiccup was due.

After almost three hours we landed at Neumayer and I felt as if I was on another planet. As at Novo, the temperature was mild and in front of me lay a seemingly never-ending sheet of white ice, but this time there were no rocky peaks anywhere to be seen. With such a flat expanse and in an environment where the air was so pure, I could see as far as the curvature of the Earth allowed me to. The weather was overcast and very calm, but still so blinding to my eyes that I needed to wear heavily tinted sunglasses, so dark that they were classed as illegal if used for driving back home. Only a faint horizontal line in the distance represented the border between sky and ice.

For almost two years I'd been looking at photographs of this place trying to contemplate if the station and the landscape around it actually existed. I couldn't believe I was finally here. As I rounded the nose of the plane, there she was, sitting on an elevated mound: Neumayer III Research Station. A real-life Antarctic castle. The station was made of modern metal panels, painted in three horizontal stripes, blue, white and red. A look-out point on the roof for any invading explorers was surrounded by a tangle of radio aerials. On the ground, snowmobiles and snowcats were parked in grid formation. The snow had been groomed and felt solid, and everything, from the landscape to the man-made elements, was pristine. It was immediately obvious that Neumayer was well looked after.

Twenty or so blue shipping containers broke the horizon as they sat dormant on the snow, spaced equally in two lines, parallel to the station. Giant footprints led away from a large white satellite dome that sat on four thick steel legs, looking like astronauts had landed and headed off to explore. As I looked around, I felt as if I was dreaming.

Resembling a small post office, the main entrance porch was red with a white roof. Draped across the entrance door between two thick bamboo poles was a white bed sheet. '*Herzlich willkommen!*', which translated simply means 'Welcome!', had been handwritten in permanent marker in large blue and red block capitals, in keeping with the station's colour scheme. The existing overwintering team from whom we would take over had made sure they were prepared, greeting us in the most welcoming fashion with flutes of champagne. I wasn't sure but it wouldn't have surprised me if it was tradition when welcoming the new overwintering team. It was like a handing-over-of-the-baton ceremony.

For a group who had been living at Neumayer for a year without a break, they looked remarkably well and happy to see new people. Their faces were tanned. The men had suitably grizzled beards, quintessential for a stereotypical Antarctic explorer, and the girls' hair was long. I didn't know what I expected, but seeing them still smiling after such a long, arduous stay, which had included their own winter of isolation, was encouraging. The standard duration for a core overwintering crew at Neumayer was fourteen months and as they had arrived exactly one year before I had, they still had a further two months to go.

Looking straight up at the windows of the station, I was desperate to get inside and see what our accommodation for the next year was like, but I was equally keen to spot my first emperor. Even though I had no idea where anything was, I stepped away from the chatter of excited voices to scour the vast landscape in the hope that one might be visible from the front door. Within seconds I saw a penguin, then another one, then lots. Emperors dotted in two long lines leading into the distance. But why such consistent gaps between them and why were they so keen on the runway? We'd not long before landed a large, loud aircraft on it. Out of overenthusiasm or tiredness, the mirage of black flags marking out the position of the runway had played their first trick on my eyes.

Neumayer III Station, named after the German explorer Georg Von Neumayer, sat at just below 70 degrees south on the

floating Ekström Ice Shelf along the northern coastline of
Queen Maud Land. The definition of 'ice shelf' had always
confused me but put simply, it's a thick platform of ice that
forms where a glacier flows down to a coastline and onto the
ocean's surface. Standing on such a large one, it felt like solid
ground. Having first opened in 2009, Neumayer III looked like
a huge ship on stilts. Underneath, enormous platforms, to
which the sixteen stilts were connected, lay flat on the ice in a
huge rectangular hole that had been dug into the ice. In between
the platforms was Neumayer's impressive collection of machinery; the garage appeared more like a warehouse in a ski resort
than an Antarctic research station. Two Toyota Hilux pickups
sat side by side, both equipped with winches, huge snow tyres
and a small crane lift. There were over fifteen snowcats, or
'Bullys' as we called them, the same kind of machinery used to
groom ski slopes. Basically bulldozers for snow. There was also
a yellow crane, a cherry picker on caterpillar tracks and eighteen blue skidoos. Seated on a moving glacier, Neumayer III
drifted approximately 200 metres closer to the edge of the ice
shelf every year, so when the building was opened it was given
a lifespan of approximately thirty years. If the ice shelf continues to move at that rate Neumayer might be at risk of breaking
away on an iceberg, so the plan is to disassemble and move it
off the Ekström Ice Shelf before then. No doubt an even more
high-tech modern construction will then take its place further
away from the open ocean.

With the station leader keen to give our group a guided tour,
we finally made it inside. Under the welcome banner I entered
the station through what looked like a thick refrigerator door.
On the first floor was a boot room filled with lines of pegs like

my old school cloakroom, only polar clothing and huge snow boots had replaced blazers and sandals. A shiny steel kitchen lay opposite with utensils hanging in neat lines from the air vents. Over a counter I could see a dining room with three long tables. There was an ice cream fridge in the corner, a shelving unit with an endless supply of sweets and at the end of the room, two enormous glossy photographs dominated the wall. One showed a bright aurora australis illuminating the sky over the station, and the other an emperor and her young chick. 'That's why I've travelled here!' I said out loud.

In the living room there were two long sofas, a full bookcase, a pool table and an incredibly well-stocked bar. I should have known, with the reputation that Germans have for their love of beer. The large living room, which spanned the full width of the station, was situated at the southern end and tall windows revealed never-ending views across the ice. During the summer, for just over two months, Neumayer experiences twenty-four-hour daylight where the sun skirts the horizon at midnight without dipping below it. The large windows offered the perfect place to watch the low sun as it travelled in an anti-clockwise rotation around the station. With its bedrooms, doctor's surgery, office spaces, science laboratories, a small gymnasium and even a sauna, I knew I was going to be well looked after.

That evening, with the station at full capacity, Will, Stefan and I settled down in our shared bedroom. The room had two large wooden wardrobes, a small square table with a shelving unit above and sockets, which provided either an internet connection for my laptop or a phone connection for the regular handset, but not both at the same time. Two single beds lay lengthways along one side of the room, each with a bunk above that folded down. It was tight but comfortable and apparently resembled student halls in a university, though I wouldn't know. I'd acquired the single bed nearest the door and even though the bed above me was unoccupied, like a child building a den, I pulled it down for extra cover. A single window revealed the never-ending view west across the ice sheet. Twenty-four-hour daylight was not something I'd experienced before and I was nervous about how it would affect my sleeping patterns, but when I pulled down the blackout blind, the room went dark. I had nothing to worry about.

The next day we all woke early to our first morning at Neumayer, keen to start learning the ropes of living in our new home. So keen, in fact, that we'd forgotten to alter our clocks back from South African time and arrived for breakfast an hour early. Throughout the morning it became clear how active the summer months at Neumayer were. It was a busy place, providing accommodation for up to sixty people, all with different individual duties. Mechanics ensured the transport and enormous generators that powered the station remained in good working order. Electricians and maintenance staff updated components and repaired any scientific equipment that needed attention. Chefs, doctors and IT specialists took care of everyone who needed it. The common aim was ensuring the scientific experiments continued without disruption, which was, after all, the reason the station was there in the first place.

The daily routine was well planned and everyone was incredibly well looked after: breakfast was between 6.30am and 8am, a *bratwurst* (sausage) break followed at 10am, lunch was between 1pm and 2pm, then tea and cake at 3.30pm and dinner was between 7pm and 8pm. Even though people were working for long periods away from their families in pretty difficult conditions, morale always seemed high, but with such a short four-month summer season each year, work had to be completed quickly and efficiently before the winter weather set in.

I was desperate to get outside and see my first emperor penguin; the station leader arranged to give us a tour of the surrounding ice during our first afternoon. I'd been at Neumayer less than twenty-four hours and the whole team was excited for our first penguin encounter. For Will and me, the emperors were the sole reason we had committed so much. Late in the afternoon, we met outside the station by the snowmobiles. With each machine taking two people, I hopped on one with our mechanic. I had no idea what our leader had planned or where she intended taking us but driving along the flat landscape that afternoon was my first experience away from the station. Following a line of flags, the fleet of seven skidoos sped along the snowy surface. The air was cold and I had underestimated the amount of clothing I needed to wear; my face began to sting. Within ten minutes, in what seemed like the middle of nowhere, everyone stopped. Even though we'd been hammering along at speed away from the station, I looked behind me and it seemed within easy reach. The empty landscape put a whole new perspective on distances.

Straight ahead of us, our leader pointed towards the northern horizon. Just like the flags along the airstrip, black silhouettes were dotted across the ice in the distance. 'Penguins,' she said in a monotone voice. Having spent a year living close to the colony, for her the novelty of having emperors as neighbours had clearly worn off. Squinting, I tried to focus my eyes on one point across the white expanse. I wasn't convinced and wondered if my eyes were playing tricks on me again. I felt excited and, all of a sudden, restless. If they were emperors, I wanted a better view. Being told they were emperors was no good; I needed to see them. But as quickly as we had spotted them, they spotted us. Sliding on their bellies, two emperors began to rush over. I was aware that the species was inquisitive as very few emperors have ever encountered humans, but this seemed daft. I'd never had a wild animal come speeding towards me like this and it felt an honour to be trusted. I dropped to my knees and sat back on my ankles, letting the two birds come as close as they felt comfortable. The closer they slid, the more surprised I became. The rest of the group of emperors followed the two bolder individuals but went straight past, taking no notice of us. The two adults, however, just did not stop.

I didn't dare move in case I surprised them, but as they approached almost to touching distance on their bellies, I felt maybe I should move out of their way. But just before I did, they both rose to their feet, bowed and with their signature trumpet call, introduced themselves. Being so close, the intensity of the sound went straight through me. Despite being so powerful that I could feel my ears vibrate, the sound wasn't unpleasant. It was a sound I'd heard on television and online during research for my trip, and to hear it in real life was incredible. On my knees, I was the same height as they were at well over a metre tall, and I could see straight into their eyes. The pair were so close I could see every fibre on every feather. They seemed so relaxed standing just a few feet away and I could feel their charming and peaceful personas. Sitting in such calm conditions in the short and relatively tranquil summer season meant the penguins could relax; it was hard to imagine them battling the dark, raw, seemingly never-ending winter that I'd come here to film them in.

I looked at Will, crouched next to me, in disbelief. We wanted to say things to each other, but we were totally speechless. We both chuckled, knowing what this meant to each other. I felt like the luckiest man alive and I found it hard holding back the tears. In front of me, just a few metres away, was one of the world's most famous and favourite creatures in its natural habitat.

Following a few busy days settling in, Christmas arrived and it was a day I'd weirdly looked forward to for a long time. A white Christmas was something I'd only experienced once before but even though I was in Antarctica, I didn't expect to be sitting on top of two hundred metres of the stuff! I'd always fancied being away filming over Christmas as in my mind it represented true commitment and dedication, but what it actually turned out to be was my first big test without Becky. I'd not been a huge fan of the day itself so I didn't feel it would be too much of a problem, but with eleven months ahead of me, the realisation was still sinking in. Will also had mixed feelings and, being one of only two Englishmen on the station, he made sure some traditions were maintained. German nationals celebrate Christmas on 24 December, so still with no camera equipment, we sat down to a Christmas banquet of duck with the other sixty people on the station. The operations manager made a speech and emphasised how Antarctica was still a place of freedom where nations unite and support each other. With representatives from Germany, Hungary, Sweden, Switzerland, Russia, South Africa and of course the United Kingdom in the room, it was a poignant moment and being in such company reminded me what a privileged position I was in.

Late the following morning I received the best Christmas present I could have asked for: delivery of all our camera kit from Novo. Each day since arriving I'd been told it was on its way, but each day that it hadn't arrived, I was beginning to wonder if I shouldn't have trusted Chuck after all. At 1.30pm, a tiny distant plane appeared in the blue sky. With a small propeller on each wing it was a fraction of the size of the one I'd boarded just a few days earlier. At last, we had some equipment. Unbeknown to me, secreted in a couple of the cases were Christmas presents and in the early evening, Will led Stefan and me up onto the roof of the station. To the south, the sun hung low and the ice glowed orange. An upturned cardboard box draped in a red towel was sitting on the roof. Will had decorated it with a miniature Christmas tree and hung small felt penguin decorations from each limb. Underneath were presents wrapped in emperor chick paper and a card addressed 'Team Penguin'. Passing each of us a beer, Will knew how to lift the spirits.

In the new year, *Polarstern*, a 118-metre-long German icebreaker ship, arrived at the edge of the ice shelf. Having made its way through ice up to three metres thick, it was greeted twelve kilometres from the station by a fleet of snow-cats ready to transport its cargo back to the station. It had sailed all the way from Germany, stopping only in Cape Town to collect fresh food. On board was a huge crew of scientists, chefs, mechanics and pilots. The ship carried not only pilots to guide it through thick sea ice, but also specialist pilots to fly the helicopter that had travelled with it on the helipad at the rear.

As well as transporting supplies for the station, *Polarstern* had arrived with a lot of our extra camera equipment that we hadn't needed to fly with. Being away from home for such a long time I'd also been allowed to fill two large aluminium boxes with personal items to help pass long periods of inactivity during storms and the extended winter darkness. Having been warned that the weather would become unpredictable, especially in the winter, and that unsettled conditions could last for days if not weeks on end, I'd decided it would be a good opportunity to learn some new skills. In one box I'd packed the standard items, a framed photo of Becky, my mum, dad and sister, board games, a pack of cards and some of my favourite books. Two cases of fly-tying materials would give me an opportunity to replenish my fly fishing boxes. Over the summer, before leaving the UK, a good friend had kindly made some felted models of Willow and Ivy, as well as a miniature version of my local pub to remind me of luxuries I wouldn't have access to.

In the other box, however, I'd gone left-field. I'd crammed in a unicycle (broken down into seat, wheel and pedals) and some GCSE German books. Having regrettably paid little attention during German lessons at school, this was going to be my best opportunity to learn a foreign language, as I'd be surrounded by German nationals. I'd secretly hidden three golf clubs in a camera tripod tube, so I'd also packed some orange air balls and a square foot of artificial turf, which were nestled in amongst the unicycle's spokes. As a proper Englishman, one home luxury I knew I couldn't go without was Earl Grey tea, so twenty packets of loose leaf had also been thrown in, along with my favourite tea strainer and a china mug, heavily packaged in bubble wrap. Finally, after I expressed interest in learning a musical instrument, my dad had bought me a brand-new violin, which I foolishly believed that I could teach myself how to play! No matter how long the storms were, I was prepared.

With an enormous crane positioned on the front of the ship, *Polarstern* transferred container after container onto the ice. Snowcats and specialist sledges with huge iron skis queued up patiently. Towing sometimes four or more at a time, the route back to the station across the ice resembled a busy train track. To support the large number of personnel over the busy summer and the team of twelve of which I would be a part through the eight-month winter, the food delivery included 1,500 litres of milk, 900 kilograms of potatoes and 5,000 eggs! A large hydraulic lift incorporated into the centre of the station made transferring heavy loads between floors light work.

One storeroom, kept at 5 degrees Celsius, was filled with jars and bottles, vegetables and individually placed pieces of fruit. I watched the chef carefully position apples and pears one by one in neat lines and I wondered if it would actually prolong their life. The freezer room opposite was kept at a constant −20 degrees, which required full winter clothing when inside while we helped restock. Boxes of frozen vegetables, litres of ice cream and a lot of sausages were to keep us fed through the isolated months. Helping restock shelves for a couple of hours gave me my first taste of what filming outside would feel like.

After only a few days, *Polarstern*'s job delivering supplies to Neumayer had been done and the ship could finally continue around the coast of Antarctica to commence its own scientific research while the weather allowed. It was a foggy but very calm evening at the North-Eastern Pier with light snowfall and, as was traditional, all the people staying at Neumayer gathered to wave the ship off. As all the crew lined themselves up along the ship's starboard side to wave back, the tannoy that was usually used for the captain to communicate to passengers came to life. At full volume, Andrea Bocelli's 'Time to Say Goodbye' blared out across the frozen landscape. It was very surreal. I looked around at everybody else waving, struggling to believe what was happening. Ironically, sea ice had drifted in and surrounded *Polarstern*, preventing it from actually leaving for another two days. It turned out 'Time to Say Goodbye' was played prematurely.

During the first few weeks of settling in, I'd been keen to show the station staff we were part of the team, not just a film crew that had turned up to take advantage. Out of the large number of people, I was worried there could have been a few who weren't comfortable with us being there. Everybody had their own jobs and I was anxious that there was a possibility people would get a little jealous of us heading down to the penguins every day. It was one of the few perks of working there, after all! As a crew, I felt we had to be sensitive. It was our job to make a film about the penguins but there was more to the trip than that and maintaining relationships was equally, if not more, important. Sticking around the station, I built the cameras and arranged our kit room. Generously, we'd been given free rein of a large space next to the generator room. Despite being loud, it was lovely and warm and having such a large amount of space allowed me to arrange everything and know exactly where each piece of equipment was. We'd travelled with an enormous array of kit and with no options to replace anything that broke during the winter, we had two of almost everything.

Before I could begin filming I had to work out a way of transporting the camera gear safely to and from the penguin colony. Using heavy-duty ratchet straps, I secured a couple of large empty aluminium boxes to each Nansen sledge. The sledges were long, wooden and rigid and hooked onto the tow hook on the back of each snowmobile. Out of the eighteen skidoos, I adopted number 10, and rather than attaching the box containing our expensive camera to the sledge, I sat it on the metal platform behind my seat. It was still bouncy but the skidoo's suspension made sure that, so long as I packed the camera tight in its box, it wouldn't rattle into pieces. It was all a massive learning process. We rehearsed travelling in all types of weather and ran through the assembly of camera gear in snowy conditions. It felt silly as the weather was still relatively good, but necessary as we knew it would change.

The summer season was in full flow. I had a storeroom full of camera equipment, a warm comfy bed and a skidoo weighed down with my filming kit. Weather conditions remained settled throughout the first couple of weeks of the summer and although temperatures averaged between −2 and −5, I was forced to continuously apply heavy factor 50 sun cream no matter how overcast the sky was. Spending all my time on a land of snow and ice, which reflected almost all of the sun's ultraviolet rays, I was effectively exposed to double the amount of sunlight. Only twice did I notice the temperature rise above the zero mark, during which time the snow outside became thick and sugary, making walking tiring and travelling by skidoo challenging. More than once I rolled my snowmobile over as I ploughed its skis into deep snow, but thankfully not at speed. Normal clothing would suffice for the majority of the time. A thick jumper, trousers and snow boots kept me just right. A hat and gloves weren't necessary a lot of the time but the one item I couldn't do without were my dark sunglasses. The landscape, even during overcast days, was blinding and no matter how long I gave my eyes, they couldn't adjust. It was dazzling and painful and at times gave me headaches. As time went by, however, I adjusted to life in Antarctica. Even though I thought about Becky and home all the time, I felt settled, calm, and I was ready.

2.

The Emperors Return

On 1 February, the sun touched the horizon for the first time in two months. Daylight hours had started to shorten, but the sun was setting for only a short period over midnight and there was still no sign of any darkness. I'd been living at Neumayer a month and a half and already it felt like home. Despite having to film around the weather, a daily routine at the station had started to take shape. It was still at full capacity with almost sixty people working around the clock, but it hadn't taken me long to develop friendships.

When I'd arrived at Neumayer I'd been thrown into a very busy summer season with everybody having their own jobs to be done. I, on the other hand, had felt lost; not being a scientist or an employee of the AWI, I had felt like a bit of an outsider, but as time had gone on I'd settled down. Even though my job was totally different to anybody else's on the station, I started to feel part of the team. I was beginning to feel comfortable and settled, but unbeknown to me, I was about to face my first big psychological challenge.

In previous years the sea ice covering Atka Bay hadn't broken up, even through the warmer months of summer. Along the northern exposed edge, two enormous tabular icebergs sat dormant, so large that their bases rested on the ocean floor deep under the water. I wondered whether the icebergs had an influence on the stability of the ice sheet covering the bay, giving it just that little bit more protection and strength. But this year was different.

We'd made the risky assumption that we would have access to the surface of the frozen ocean from the moment we arrived through the whole duration of our stay, but to my astonishment and horror, overnight the whole area of ice covering Atka Bay shattered and broke up. Over one hundred square kilometres of ice was gone in just a matter of hours. This not only meant the penguins couldn't return until it refroze but also that we wouldn't be able to get down closer and onto their level to film them. When the sea ice had coated the bay for the whole year the emperors had endless time to breed and raise their chicks, but this year was a different story. By early February the entire bay was open water, with pack ice and floating icebergs drifting in and out, but most importantly for us there was nowhere for the penguins, or us for that matter, to go. We were stuck on top of an ice shelf fifteen metres high at the top of a cliff with no penguins to film.

The initial plan had been to start filming as soon as we got to Antarctica, to get a feel for the behaviour and characteristics of the birds, but with such a dramatic and unexpected change I hadn't been given much of a chance. Following the emperors' relentless annual breeding process each summer, they spent weeks at sea feeding, gaining weight and recovering in preparation for the next season. I was expecting at least a few weeks without the birds, but I didn't expect to be without them for months. The ice on which they'd bred had fragmented and floated off, and they had no option other than to go with it. Suddenly, the thought of having nothing to film while the emperors were away terrified me. What was I going to do?

With no penguins in the area, I potentially had a few months to fill while I waited for them to return. Despite the film not requiring much footage of drifting ice floes and icebergs, filming from the top of the cliff felt as if I was maximising my time. Standing at the edge, I looked down towards the open ocean below me with the slight tension of a long rope pulling against my waist. The harness I was wearing squeezed tightly around my thighs and metal carabiners jingled against each other every time I moved my feet. Being in such a precarious position at the top of the unpredictable cliff of ice, I made sure I was connected to the safety of my skidoo at all times. Parked further back and weighing in at almost a quarter of a tonne, it was my anchor in case the ice gave way. Plunging into water that hovered around zero degrees Celsius was not something I wanted to experience. A separate length of rope kept my camera safe. Water crashing against the base of the ice cliff eroded deep horizontal cavities and every now and then, with the addition of warmth produced by the sun, large lumps of ice would tumble down into the water. It was a beautiful sight looking down into the clear blue depths, but potentially a very dangerous place to be.

Although I was in such an enormous, wide-open land-scape, movement across the ice was strictly controlled. This was to keep everybody safe. An ice shelf is essentially a glacier and can develop fissures and cracks at any time anywhere across its vast platform. This being the case there were only three routes that led away from Neumayer, all of them in straight lines heading in a northerly direction towards open water. The area of ice these routes ran along was heavily monitored by station staff to make sure no cracks appeared and that they remained safe areas to use. Each route was loaded onto the enormous collection of GPS units that Neumayer owned, including the ones mounted on every vehicle. Every mode of transport the station owned had a GPS on its dashboard.

The three routes were marked with flags, two- to three-metre-high thick bamboo poles standing in deeply drilled holes in the surface of the ice. Tied onto the top of each pole, red, blue and black squares of fabric wafted in the breeze. To make travel safer during times of bad visibility, every hundred metres a flag stuck out of the ice, placed in immaculately straight lines. The westernmost route led to the edge of the ice shelf known as the Northern Pier. The central route terminated at the North-Eastern Pier and a final shorter track led to *Pingirampe*, translated simply as 'penguin ramp', the closest portion of ice shelf to the emperor colony. Over the years these areas along the shelf edge had been the safest places for cargo ships to dock and unload supplies for the station, hence the term 'pier', and with the routes having been heavily trafficked by snowcats, tracks remained in the snow from when *Polarstern* had docked against the North-Eastern Pier over a month before. Getting around on just three different roads didn't take too long to memorise and actually made life much easier.

Floating on the surface of the ocean, ice shelves are generally flat, featureless landscapes and looking across our ice shelf every day I started to worry. Despite being in the one place on the planet I'd dreamt of reaching, Antarctica's landscape looked extremely uninteresting and from a film-maker's point of view, I wondered how on earth I was going to make it look visually exciting. Even though it was still February, my Antarctic adventure was already the longest film shoot I'd ever been on but I'd barely filmed a thing, and with only a flat expanse of ice and open water to focus on I found it impossible to visualise the place in winter. The positioning of Atka Bay, however, was one aspect that had made Neumayer an attractive option when looking for a location for making a film about emperor penguins. Films documenting the emperors' life cycle had been made before, but to make our film stand out, the surrounding scenery and landscape was an important factor.

Previous films, including *Planet Earth*, had been filmed at colonies that lay in amongst clumps of enormous icebergs, locked in place by the frozen ocean on which they floated, and this had given each image in those films grandeur, drama and scale. On such a flat landscape, however, as the sole camera operator, I would be forced to give the film more style, to bring each image to life by focussing on the penguins themselves rather than the visually uninteresting landscape behind them. Selfishly, however, I was desperate for icebergs. These enormous irregular structures of frozen water symbolised

Antarctica in my mind, and not having any within sight was difficult to accept. I began to feel as if the film would be a failure without them.

Every day, new icebergs would appear on the horizon drifting from east to west with the ocean's currents around the continent. I never had any idea how far away they were and with nothing to compare against for scale, I had no way of appreciating their size. If I was lucky the ocean's swell would divert a few into the bay, sometimes just a lone berg, but more often than not, clusters would travel together. The eclectic mix of shapes and sizes fascinated me and I wondered where their lives had begun. Large pure-white structures with a flat surface would indicate they had broken away from an ice shelf somewhere else on the continent, just like the one I was standing on. In some cases where the icebergs had tipped slightly they revealed a smooth sky-blue underside. Deep, dark caves and large holes through which the sun shone weren't uncommon. Electric-blue icebergs occasionally appeared, indicating ice that hadn't seen light or air since its formation potentially thousands of years before. It reminded me of looking up at an old oak tree back home, trying to imagine what had happened on the planet since it had begun its life. Only this was on a grander scale. Unlike a tree that may have seen several different kings and queens reign, the ice could have been around during the time of the woolly mammoths. Scientists in Antarctica recently drilled a core of ice believed to be 2.7 million years old!

As these enormous structures drifted into Atka Bay I longed for the ocean to freeze solid and lock them in place. Not only would it give me my backdrop to film the penguins in front of, it would give me somewhere to explore when the sea finally froze. Temperatures still hadn't dropped consistently enough to freeze the sea and I knew I had a while to wait. Watching them from my elevated position at the top of the ice cliff, I couldn't tell how close they actually were, but I could see they reached much higher than the ice shelf I was standing on.

Surrounding the base of each iceberg was pack ice, the thin layer of sea ice from the previous winter that had broken up but hadn't melted. Smashed into chunks by the ocean's waves and currents, it hid almost all of the water's surface at times and into the distant horizon small, sharp, jagged ridges popped up out of the mangle of ice. On calm, bright days, I watched as pack ice drifted past and I dreamt of hopping from one piece to another like stepping stones. How far could I have travelled? Within the space of just half an hour, however, the bay could go from being packed solid with ice to completely open water with not a floe to be seen. It amazed me how quickly the ocean's appearance changed and even watching distant icebergs drift across the horizon, I couldn't believe how quickly structures that weighed tens of millions of tonnes could move. The sea's surface seemed alive yet it was under the complete control of the weather and currents.

As well as the emperors, Antarctica is home to a huge array of wildlife that is the stuff of dreams for enthusiasts across the world. Some of these incredible species can't be seen anywhere else on Earth, but I had been continually told that around Atka Bay, apart from the emperors, there wasn't a great deal of other wildlife to see and it worried me. While on the longest couple of filming shoots I'd done prior to Antarctica, I'd ended up almost losing interest towards the ends of the trips. Having just focussed on one species for weeks on end, I had become bored and needed variety. I was nervous about the prospect of only having one species to film; having other wildlife to watch would have kept me engaged.

I knew there wasn't a chance of this during the winter as nothing but the penguins could cope with the severity of the weather, but I found it hard to believe there wasn't anything else during the summer. With no penguins for a couple of months while they spent their time feeding, I wondered how I'd cope. I'd never gone a day at home without stopping and spending five minutes watching a wild bird or an insect go about its daily life. The prospect of seeing nothing was daunting. I'd quizzed station staff back in Germany prior to leaving about what to expect, but as wildlife wasn't a main point of focus for Neumayer they weren't sure what would be about. Stefan, during his previous time at Neumayer, had become frustrated by not being able to get outside as much as he had wanted due to work commitments, so he couldn't answer my questions either. It was a case of discovering what we could, and by being outside on the ice all day every day, we stood a good chance of seeing any wildlife that might be there.

The Northern Pier was the furthest point on the ice shelf from Neumayer and it quickly became my favourite place during summer. It felt remote and being almost an hour's skidoo drive away it didn't get many visits from anyone else, so I knew I could find peace there. As with everywhere else, the station appeared only a stone's throw away looking back at it in the distance, but in reality I had travelled over twenty kilometres. I'd overcome the frustration caused by the sea ice having broken up; as it was out of my control I had to accept it was just one of those things. But on the other hand, open water presented a habitat I hadn't expected to encounter. The environment below the water's surface had in previous years been cut off to any terrestrial life by an impenetrable layer of ice, but suddenly now it had become accessible.

As I looked down from the edge of the cliff at the Northern Pier, ghostly snow petrels, a species of Antarctic gull, floated in the air. So white and pure, only their black eyes and bills gave them away against the ice as they drifted along the edge using the rising sea breeze to guide them. Just like the penguins, these birds were recovering from a busy breeding season, feeding in patches of open water as they travelled north when the temperatures began to drop. Atka Bay, without mountainous rocky outcrops, was a far cry from their inland nesting territory. Having been spotted as far down as the South Pole, snow petrels are the second most southerly breeding bird in the world, after the emperors, but like any sensible species, they made their retreat while the weather still allowed.

Antarctic petrels, a slightly larger version of their snowy cousins with a light brown mottling along their wings and head, joined in cruising the air along the ice cliff. Folding their wings, they swooped, dropping down to land on large ice floes. A few hundred metres out to sea from the Northern Pier, the cliff of one of the two grounded icebergs towered out of the ocean, dwarfing the ice shelf. Against its gigantic overhanging walls, the dark silhouette of a giant petrel scythed through the air. With a wingspan of over two metres, solitary birds patrolled patches of open water looking for opportunities to feed. As it flew closer, I could see that it was one of the biggest birds I'd ever set eyes on and as quickly as it had approached it glided round the sharp edge of the iceberg and disappeared from view.

High above me, the familiar high-pitched trill of twelve Arctic terns sounded. Even though I was at the other end of the planet to the Arctic, I recognised their call immediately. I wondered if these birds had come from my home; their migration can see them travel almost from pole to pole and back each year. I'd spent a lot of time up close filming nesting Arctic terns on the Farne Islands back in the North-East in the UK and knowing how fragile and small they were made the journeys they had undertaken seem nothing short of a miracle. Their trip down to Neumayer must have been far more arduous than mine and as they elegantly flew off into the distance I wished them luck on their return journey north.

Although I'd been told to expect nothing other than emperors, Atka Bay was proving to be an Antarctic oasis. Although the birds we were seeing were so special, the small variety meant the novelty soon wore off. Back home, my environment was so rich in wildlife that I was used to seeing many different species every day; in fact I could guarantee more species on my garden bird feeders at any one time than I was seeing in Antarctica. Despite the privileged views I was experiencing, I missed the variety. Unbelievably, I was missing common birds such as blue tits and house sparrows, which I'd taken for granted back home.

Against a light-blue background of submerged ice I could see dark silhouettes of large shoals of fish moving through the water. The reason the emperors, and in fact everything else I was seeing, are present around Atka Bay is due to how rich the seas are. The Southern Ocean covers only 5 per cent of the world's seas, yet it accounts for over 20 per cent of the world's sea life. Life in Antarctic waters is so abundant that the ocean contains up to six times more life per square metre than other oceans. For such a cold and desolate place *above* the water, I was amazed by how healthy the ocean seemed to be. Incredibly, long hours of sunlight promote algae growth during summer months, frequent turbulent waters maintain a constant supply of nutrients around the surface and the cold temperatures can hold more dissolved oxygen than warmer waters. With such ideal conditions, as many as 35,000 krill (tiny shrimp-like crustaceans) can swarm in one cubic metre of sea water, providing food either directly or indirectly for pretty much all the other animals in Antarctica. Everything relies on the sea.

The long summer days on top of the ice shelf watching the conditions on the ocean change were so peaceful. There were still no penguins but I knew they were out there somewhere. Despite the weather being consistently settled, each day felt different. Temperatures were comfortable, nothing lower than −8, and it surprised me how quickly my body had adapted to what I'd usually have classed as severe back home. I was still managing without hat or gloves for the majority of the time, only donning them for journeys on the snowmobile, and with the sun still shining down from high in the sky, the bright, blinding appearance of the landscape was maintained. I still found myself reapplying sunscreen every hour to my exposed skin and balm to my dry lips. The sun was gradually getting lower in the sky each day as autumn approached, but I barely noticed.

Deciding on a change of scenery one day, Stefan, Will and I headed back to the North-Eastern Pier where *Polarstern* had docked just over a month before. At its tip I could see the Northern Pier to my left just a few hundred metres across a small inlet in the ice shelf and to my right the long winding edge of ice led along the western portion of the bay. The snow petrels seemed to be following me as they patrolled up and down the cliff. One had even glided parallel to my skidoo for almost half of my fifteen-kilometre journey to get to the pier. In front of me, 'bergy bits', or large chunks of ice that had broken away from icebergs, lay floating on the serene, seamless surface of the pristine water. It was incredibly calm and there wasn't a ripple to be seen across the open ocean. Any disturbance in the tranquil water would have stood out for miles. Sitting on the seat of my skidoo with both legs to one side, I looked out over the water. The cliff of ice was around twenty metres high and, not yet being roped up, I was a good distance away from the edge.

All of a sudden a loud 'whoosh' sound rose from below, beneath the wall of ice. I looked at Stefan and Will. 'What on earth was that?' I said as we rose to stand higher for a better view. Ripples slowly flattened out across the calm water as they appeared from behind a sharp edge. I wondered if a large portion of ice had tumbled down into the water out of view. For a few moments, there was silence. Rather than rushing about to get rope and harnesses ready, we turned our heads, restlessly scanning every square inch of ocean in front of us. Suddenly, like a bow wave, the head of a whale pierced the surface and exhaled. It was a minke whale. As it broke the surface to replenish its oxygen, it was unmistakable. Behind it, another followed. Both emerging twice, they each took one last huge breath, arched their backs and dived. Their small, insignificant dorsal fins at the ends of their backs slipped down through the surface.

It was the first view of what would turn out to be daily sightings of the minkes, and it felt incredibly special. Often travelling in pairs, they worked up and down the edge of the ice shelf travelling around the perimeter of the bay and feeding tight against the cliff wall. I wondered if Atka Bay was a favourite old feeding ground that they'd been unable to visit due to the ice for a few years. Watching them in what looked like inviting waters, I felt like jumping in and swimming alongside them. The view they were getting under the water must have been incredible. On occasions a huge silent shadow of an individual cruised past just below the surface. The clarity of the water was exceptional but only the grey shade of the whale's skin against the deep blue waters revealed its presence. Despite seeing them every day, each sighting felt special and exciting. There's something about seeing cetaceans that has always thrilled me. The fact that they live in a world so separate from ours makes those occasions when we meet even more special.

Later that day I peered through my binoculars, scanning the skyline. A bank of cloud had shrouded the ocean; currents had pushed in, forcing the ice tight against the cliff, and despite there still being no wind it seemed a completely different day to the morning. The broken pack ice on the horizon was highlighted against a dark grey band in the sky. Long stretches of dark-blue open water reflected against the low clouds, creating an inverted map of the ice patterns on the sea. It was quiet and there wasn't a bird in sight. Seeing the landscape through my binoculars compressed the layers of ice and gave the flat landscape more contrast. Everything became clearer. As well as reflecting against the clouds, the open water reflected upon the edge of each floe, turning the ice grey. It was irregular and heavily eroded, and the messy ridges gave it a cluttered appearance. It was like floating debris. However, there wasn't an iceberg in sight.

In the corner of my eye a dark shadow ran across in front of an icy ridge. Then another. It was distant, but the motion against the static appearance of the floating ice lured my eyes to focus in. From behind a tall white lump, another shadow lowered diagonally, disappearing between two floes of ice. Then another.

'Orca!' I shouted, pointing into the distance. Scrambling the camera kit, I searched the horizon through my long lens. The ocean appeared locked tight with pack ice but each time one surfaced between ice floes it revealed an area of open water invisible from my viewpoint. As the ice drifted, large patches of water became more obvious and my elevated position enabled me to see the gaps. I looked through my viewfinder, panning my lens from left to right, but the distant dorsal fins had disappeared. Where had they gone? I lifted my eyes and looked down below the ice cliff. The areas of open water were disappearing as drifting ice started to build up once again.

Orcas are known for their ability to hold their breath for up to fifteen minutes and I began to believe they'd taken on board the necessary oxygen and dived deep, heading away from us. Half an hour went by; nothing. They must have surfaced but I couldn't see where. I turned the camera off to save battery. Will stood tall on a small mound next to me, desperately scouring the ocean. He'd never seen an orca before and was keen to pin them down again. Stefan stood the other side with his own camera in hand. He, too, had never seen one and was equally keen. I looked at them both.

'What do you reckon? Should we pack up?' I asked. I was sure the orcas had moved on, around the coastline and out of view. Will gave it one last look, slowly scanning with his binoculars from left to right, but nothing. Then all of a sudden, 'Here, right here. Oh my word!' he shouted, pointing his right hand down almost towards the base of the cliff. From a seemingly non-existent gap in between two floes, water and slushy ice erupted. From the bubbling explosion, an orca's head appeared, rising vertically out of the water. Nose, then mouth, and eventually its whole head. The whale's tiny eye made contact with us and within a couple of seconds it lowered back down into the water. The gap in the ice was immediately plugged with a thick layer of slush. But as soon as it had disappeared an explosion of air rose from below, parting the ice on the surface. We were being 'spy-hopped', a behaviour that involves orcas and some other species of cetaceans poking their heads out of the water vertically to get a better view of the surrounding area. In between more floes, more whales appeared. The whole pod were bobbing up and down in front of us.

I'd been lucky enough to see orcas before in the UK around the shores of Shetland, and in Canada, but never this close and never doing anything other than surfacing for air while travelling. The scene was exactly the image I had in my head of what I had imagined Antarctica to look like, but capturing as much as I could on camera prevented me from being able to fully enjoy the moment. All I wanted to do was watch, but I had to catch what was happening on film. With there being four different types of orca present in Antarctic waters, all with different feeding habits, I wondered if they had preyed on a few penguins during their lives. It was a weird but exhilarating feeling being watched for over an hour by a huge number of whales.

Orcas are mammals that need to surface to breathe, so a frozen layer of sea ice would effectively drown them. As pack ice drifted in and out, the orcas were checking for gaps where they could get their next breath of air on their way out of the bay. This in turn helped them to establish a route back out to safer open water. As well as looking for breathing holes it was obvious they were checking us out; we must have stood out for miles in our bright red polar suits standing on top of the ice cliff looking down on them. Moments after the orcas stopped surfacing, a few ice floes parted, forming a small area for them to breathe. Gambling on the orcas using it, I pointed the camera in its direction, locked the tripod in position and let go. One after the other, males, females and young individuals powered through, the wake on the surface of the water slamming into the ice floe that they were diving beneath. I watched through my viewfinder and started counting. Ten, twenty, thirty, and still they charged through with intent. The water settled and the gap in the ice closed up. In total I counted at least sixty-nine. I had no idea there were that many and I was speechless. Inspecting for more gaps in the ice in the distance, I waited, but nothing. Unknown to me at the time, that would be the last occasion I'd see orcas in Antarctica.

Back at the station, activity was slowing down. We were into the second half of February and temperatures still hadn't noticeably dropped. The weather had been a lot more settled than I'd expected and even though the sun was setting for longer periods each night, twenty-four-hour daylight still dominated, a bright twilight still illuminating the landscape for a few hours over midnight. I still hadn't experienced the danger of an Antarctic storm and it was difficult envisaging such rough weather and what damage it could do without actually experiencing the conditions first-hand. I was thoroughly enjoying the comfortable, bright conditions but part of me was desperate to see how much the weather could deteriorate. I wanted a short storm to pass over just to give a taste of how brutal things could get.

Over the following days planes landed, refuelled and took off along Neumayer's ice runway. People from all over Queen Maud Land were fleeing, just like the petrels and terns. Over the summer I'd forged some great friendships with staff at Neumayer and as they began to leave it proved emotionally testing. For over two months I'd lived alongside them, eating and socialising with and in some cases relying on them. Very quickly they were disappearing. Preparation for Neumayer's winter lockdown had well and truly begun. Snowcats that had spent the summer shunting snow and towing containers were being transferred from their parking lots outside the front of the station to their winter *hibernaculum* in the garage underneath. Parked within inches of each other, twenty Bullys filled half the space, all with batteries removed to be kept in warmer conditions during the extreme winter temperatures.

The shipping containers that housed all the summer supplies next to the station were towed four kilometres away to the winter storage area, two long lines of almost a hundred containers positioned at equal distances apart. Everything seemed rushed but Neumayer was slowing down. With only around thirty people left and unsettled weather conditions approaching Queen Maud Land, one final plane landed on 25 February to take the remaining summer staff out. It was forty-eight hours earlier than scheduled due to incoming weather and with decisions being made at the last minute, it came as a bit of a shock. I felt excited but nervous and wasn't sure I was ready to be left alone. At exactly 4pm we said our goodbyes and the door to the plane closed. The twelve of us remaining stood by the side of the runway on a sledge connected to one of only two Bullys that hadn't been put away. With all waving, the aircraft took off and disappeared into the clouds. We were isolated, cut off from the rest of the world. High-fiving, we cheered in excitement. It was as if this was the moment we'd all been waiting for. If I still had any doubts about committing to this trip, this was the time I had to put them aside.

Walking back up towards the station I stopped and looked up at it, my new home. Together with a doctor, a mechanic, an electrician, a chef, an IT specialist, four scientists, a camera assistant and a television director, had I really been trusted to look after this place for the next eight months?

Immediately gathering in the living room, we each grabbed a beer. Dr Tim announced, 'As the new base commander I want to discuss some new rules,' with a huge grin across his face. He didn't have any new rules, he just wanted us all to know his new title. He seemed very excited that we finally had the station to ourselves.

I'd first met Tim in Germany the year before. The overwintering crew of nine people (other than Stefan, Will and myself) who would man the station during the months of isolation had been selected through an interview process, and Tim had been offered the doctor's job, doubling up as the man in charge. Tim was in his late thirties. He was tall, strong and had by late February already allowed short stubble around his chin and upper lip to become the makings of an impressive beard. One of my first impressions of Tim was that he was quite intimidating; both his size and his position of authority made me nervous. I very quickly learnt, however, what a gentle giant he was. He spoke fluent English and was keen to have a good time in Antarctica. He made it very clear from the off that he was there for anyone who needed him and was more than happy to help us where possible. If there was ever a time I needed to talk to someone, I knew I could go to Tim.

Sitting in the living room together, the overwhelming excitement of having the place to ourselves overshadowed the fact that we weren't leaving it for a while and as the chef prepared a mountain of pizza we partied like we had never partied before. I wasn't usually one for drinking, but this was different. I realised I probably wouldn't get abandoned on the bottom of the planet again, so I needed to make the most of it. Before things went too far, I quickly popped upstairs. With so many bedrooms having become unoccupied I finally had my own room opposite Will, who'd remained in his top bunk. I plugged my phone in and called Becky. She'd been shopping and had sent me a photo of a plate of fresh salad and another of Willow and Ivy.

'What's for tea?' she asked.

'We haven't run out of fresh food just yet, but I suppose it won't be long,' I replied. Even though I felt anxious about being isolated, I reminded myself it was another milestone in my journey that I had to savour.

The next morning, slightly hung-over, I looked through my new window on a view east across the ice. A howling gale and white-out conditions. The thermometer read −9. It very quickly became obvious why the last plane had left earlier than scheduled. With a day at the station in store, I thought I'd better email Miles and inform him of what I was expecting to happen in April. I still hadn't told him that Becky was due to give birth without me. It wasn't that I didn't trust Miles, I just didn't want him worrying. He'd repeatedly told me that I could pick up the phone for a chat whenever I needed. He was one of the few people who completely understood the commitment I was making by being away from home for eleven months. As it was a Sunday and I didn't want to disrupt his weekend, I could get away with sending an email rather than calling.

'Something I didn't mention when back in the UK may come as a shock, but was planned and is nothing to worry about. Becky, my wife, is pregnant,' I wrote.

I cringed as I hit the send button, picturing his face when he read it. It felt good to get it off my chest and, having waited till the last plane had left, I'd given Miles no options for getting me home. I wanted him to know there was nothing to worry about and that my time in Antarctica was going to be unaffected. He cared about us all and deep down I felt incredibly guilty dropping such a bombshell on him. It took less than twenty-four hours to receive a reply.

'Let me know if there's anything I can do to help you and Becky. You make sure you keep in touch with home now,' Miles wrote in his typical kind-hearted style.

The bad weather, which lasted a few days, gave me a chance to finally unpack all my personal belongings into my own space. Outside wasn't that unpleasant but the visibility was practically zero, preventing us from filming. Ever since arriving, my two large suitcases had remained full, stuffed under my bed to save space in our shared bedroom. Having my own room now, I personalised it a little by putting up a few photos of family and home. Folding all my clothes up and placing them in the wardrobe, I emptied both suitcases. In the bottom of one was a green sponge ball, Willow's favourite. She'd been with me when I had packed and must have dropped it in there without me noticing. I like to think she knew what she'd done but in reality I imagine she'd expected me to throw it for her. I placed it on the bookshelf. It was a nice reminder of her and Ivy every time I glanced across at it.

After no more than forty-eight hours, clear skies returned over Neumayer and, keen to get back outside, I attached my sledge to skidoo number 10 and headed for the North-Eastern Pier with Stefan. From the station I could see large icebergs poking their heads above the ice shelf in the distance. With clear air, a heavy shimmer suspended an inverted mirage of them on the horizon. The system of low pressure that had spent the previous days over Atka Bay had completely changed the layout of the floating ice on the ocean and I could see immediately it had delivered us some new icebergs. I led the way on our skidoos following the flags along the route, my GPS mounted on the handlebars directing me, as was mandatory. Tracks from the Bullys, which had spent all summer driving up and down, gave a smooth, ploughed base on which we could drive along at speed.

The journey usually took around twenty to thirty minutes but as I looked ahead, the edge of the ice shelf was approaching fast and I could see the last flag in the line. I knew that the end of the line was marked by two flags placed into the ice at an angle forming a cross, a recognisable symbol that represented danger. Not seeing the crossed flags now, I assumed the weather had displaced one, but when I glanced down at my GPS it appeared I still had a few hundred metres to go. I was confused. Surely my GPS, which I relied on so heavily, wasn't faulty? I looked up and down the edge of the ice shelf and it looked different. Puzzled, I stopped the skidoo with Stefan pulling up alongside me, equally baffled.

'What does your GPS say?' I asked him.

'Four hundred metres to go,' he replied. Surely both our units weren't faulty? Standing up to look over my skidoo's windshield, I looked out to sea. All of a sudden it dawned on me. Just a few hundred metres off the edge, there was an iceberg that looked extremely similar to what we were standing on just days before. A huge portion of the ice shelf had been calved off, and the flags marking the original route were still on it. I grabbed my binoculars; the crossed flags I'd been looking out for were clearly visible on the other side. The shelf I'd spent so much time on was now an iceberg.

Just a few weeks before, *Polarstern* had docked against this part of the shelf, skidoos had driven freely down the line of flags and Bullys had pulled up to receive cargo craned off the ship. There'd been no signs whatsoever that this was about to happen. I scanned the whole length of the North-Eastern Pier from east to west assessing the damage, still dumbfounded by what my eyes were seeing. The break in the ice was immaculate, one clean-cut curved edge over two kilometres long, as if someone had run a knife along it. Vehicle tracks in the snow from days before led over the edge as if driving into the ocean. Having been left alone less than a week before, this was a real wake-up call and I felt incredibly lucky that we hadn't tried travelling during the previous days. If I had relied on my GPS in the bad visibility I'd have driven straight over the edge.

It also dawned on me that had I been at the edge when the fissure occurred and I hadn't noticed it breaking apart, I'd have gone with it. With planes, helicopters and ships having left the continent we'd have been stranded on an iceberg with no one to rescue us. It took a while to comprehend and mad ideas of jumping over cracks and roping up and swimming to safety went through my head. The reality, however, was that nothing other than a helicopter could have saved us. Back at the station we accumulated all the GPS units and re-entered new route information, ensuring the *new* edge was in the system. The faster the sea refroze and solidified the landscape, the better.

Just a few days later we had another reality check. Rushing around at the ice edge to catch the last light one evening, we decided to relocate, packing up our camera equipment to drive our skidoos to another spot. Stefan drove off first, leaving Will and me to attach our sledges and follow behind. Having connected mine I stood over Will's sledge, directing him backwards towards the hitch. Slowly he edged his skidoo back towards me and with one leg either side of the sledge I shouted, 'Just a little bit more, that'll do. Stop.' Instead of stopping, Will accidentally accelerated backwards, knocking me clean off my feet, driving the entire machine over me and rolling the sledge. Although I lay starfished and motionless in the snow, I immediately knew that I was OK, my thick red polar suit having protected me. But for a few seconds I was so shocked I couldn't move a muscle. Looking skywards, I could hear Will running over to me. His head suddenly popped into view directly above me. He looked terrified.

'Oh my God! Oh my God! Lindz, are you OK?' I honestly think Will thought he had killed me. Luckily, I had sunk into the soft snow under the weight of the skidoo, preventing any injury whatsoever as it drove over my body. Without thinking too much about what had happened I rose to my feet, helped Will attach his sledge and drove off after Stefan. It wasn't until later that evening back at the station that we really thought about what had happened. The reality was, it had been another enormous lesson and we'd been extremely lucky. Evacuation was at this point almost impossible no matter how serious the injury. The only medical help available was back at the station. It had been a close shave but talking about it over a beer that night we couldn't help but see the funny side. Poor Will had nearly had a heart attack and it had almost been him needing medical treatment, not me.

Having had such a consistently calm summer, it seemed as if the last plane leaving had triggered the unpredictability of autumn. During the summer we'd come and gone as we pleased with our skidoos, parking them in the garage underneath. The gigantic hydraulic hatch that lifted up to allow us to drive down the ramp under the station was like something out of a Bond movie, and due to the kind weather it had remained open most of the time. But with windier conditions becoming more common and the mechanics wanting to prevent snow build-up, the hatch had been lowered, closing access to the garage. Raising and lowering the enormous hydraulic door required so much effort and time that it wasn't worth doing twice every day to let us out and back in again after filming. So to enable easy access to our skidoos, we parked three of them outside the front door. To protect them against the elements they were parked on specially designed large metal trays slightly sunk into the ice. A heavy-duty tarpaulin connected along the front edge of the tray draped over each skidoo and these were tied down with twelve fiddly fabric straps. Clearing blown snow that built up around them and reattaching fiddly straps and metal buckles with bare hands in freezing temperatures was painful work, but it quickly became a necessity and part of daily life.

It took a few days to get used to our new routine but we decided the skidoos didn't need covering over in calm weather, so it didn't prove too much of a pain to begin with. Our filming habits didn't change much; we would head to whichever pier took our fancy to look out for drifting icebergs and whales. Following an early evening session at the ice edge, I experienced my first skidoo malfunction while turning it around to head for home. No sooner had I realigned the skidoo's skis and started to accelerate than my seat slumped down and the vehicle slammed to a halt. It was definitely not right, so I jumped off to have a look.

It was bad news. The suspension had taken a clear hit, yet somehow by lifting the rear end I could balance the machine back in place, freeing up the tracks that rotated and provided acceleration. I was no mechanic, but it seemed to work. I restarted and revved the engine, but again the seat slumped and jammed as it edged forward. It was going nowhere. This was the last thing I wanted to return to the station with, especially so early into our year. It definitely didn't appear to be a quick job to fix and with heavy cloud shrouding the landscape, the weather was also against us. I reluctantly radioed back to the station. 'Leave it, we'll get it in a couple of days,' crackled an annoyed voice. I jumped off, ratchet-strapped my sledge onto the back of Will's sledge and looked at my injured vehicle.

'I can't leave it,' I said to myself. 'One last try.' Will and Stefan looked at me, both thinking I was fighting a losing battle. Again, I lifted the rear of the skidoo into position, but this time put the machine in reverse, climbed on backwards and sat on the handle-bars. With straight arms either side of my buttocks I held on and placed my left forefinger over the accelerator instead of my right thumb. Gingerly I reversed, expecting the machine to collapse once again. Success however – it held. 'It may be a long ride but we'll get it back,' I said to the others who looked on in disbelief. To begin with we all laughed, not quite believing that we were attempting to reverse a skidoo sixteen kilometres across the ice shelf, but the amusement quickly wore off. Progress was slow and my finger pressing against the throttle quickly began to cramp up. I picked up as much speed as I could but compared to a skidoo travelling in the correct direction, it was slow. Will and Stefan were very patient.

Navigating over small mounds of ice was tricky and I quickly discovered I couldn't steer quite as effectively as when going forward. Bouncing up and down on the handlebars, my backside quickly began to ache, but we were getting closer. The increasing wind started to blow snow across the surface of the ice and with the visibility having reduced to nil, we finally approached the station. The mechanic was standing by the open ramp with the controller in his hand. 'That's impressive, thanks,' he shouted as I reversed number 10 past him down into the shelter of the garage. Surprisingly, he seemed appreciative rather than annoyed. I might have broken my first skidoo but I'd gone out of my way to ensure it wasn't left out on the exposed ice during a storm. In doing so, I'd maintained an important and valuable relationship with the man who would eventually fix it, and possibly a few more.

Into early March, there were still no signs of the penguins returning and the sea ice was still as unstable and unpredictable as when it had first broken up in early January. I was becoming increasingly frustrated with the situation and with making such little filming progress; I began to wonder when it would start to happen. Will was spending more time back at the station to view footage I'd filmed, but I'd not been able to offer him anything for weeks.

The temperatures had started to drop noticeably, and the weather was also becoming more unstable. Autumn had undoubtedly begun to close in. I'd been shown the area of the bay to which the emperors were due to return to breed, and from the edge of the ice shelf I looked down onto patches of pack ice and ripples in the water. I'd become so used to looking onto an open ocean that I couldn't envisage anything else. I'd lost count of the number of beautiful icebergs that had drifted into the bay, many of which had glided over the area to where the penguins were due to return. But with no ice on the surface to freeze them in place, ocean currents had carried them back out and off over the distant horizon. The odd single penguin had started to appear in the distance, standing on drifting ice floes, but as we were expecting up to 10,000 individuals, seeing the odd isolated bird didn't mean anything.

As the sun spent almost a further fifteen minutes below the horizon each day, I began to notice temperatures plummet, especially at night when the hours of darkness had finally started to encroach. For a few weeks I hadn't noticed the sun transferring any warmth onto my skin, despite it still burning me easily. On cold, calm mornings after a clear night the bay started to become coated in a very thin layer of ice. Finally, after months of questioning if it actually would, the landscape was beginning to refreeze. On each visit we made to the edge I'd test the solidity of the sea ice by throwing a small chunk of snow down over the cliff. A quiet plop as it hit the slushy surface signified it still hadn't set.

Despite it not yet solidifying, the movement in existing ice floes and icebergs began to noticeably slow down. The bay was definitely becoming more predictable, but the tiniest bit of wind causing a swell to build in the sea would loosen it all up again. Rather than freezing at zero degrees Celsius like fresh water, saltwater requires a temperature around two degrees lower, depending on its salinity, but despite the low temperatures, Atka Bay was in need of a few days of settled conditions to properly start off the refreezing process. It just didn't seem to be settled for long enough.

As the autumn equinox passed, dawn at the edge of the ice shelf came at a more sociable hour. Getting kitted up and ready to face temperatures that had started to drop as low as –20 wasn't as difficult at 7am as it had been at 3am. Early starts are hard enough when filming wildlife but felt so different in an Antarctic environment. Early-morning mist became a common occurrence, rolling in over the water and coating everything it came into contact with in a thick layer of frost. The rising sun, despite not offering me any respite, did appear each day to burn off the fog, leaving the patches of open water smouldering as streaks of golden vapour rose from its crusty surface.

Conditions were definitely changing. For a week I'd been throwing snow over the cliff edge down onto the sea to monitor its solidity, but every time it had sunk straight through its slushy surface layer. But on 27 March, in the evening, the low golden sun shone across a thin, flat layer of ice, which appeared different. An enormous rectangular iceberg sat a few hundred metres from the edge of the ice shelf, and on the shiny surface a mirror image reflected back up at me off the ice. A thin elastic crust of nilas ice had formed on the surface. Easily bent, its pattern of interlocking fingers froze into solid ripples, giving the ice a wavy appearance. As the currents shifted the surface ice, the edges of each piece cut through its neighbouring sheet like a razor blade, splitting the membrane into fingers that shifted under and over existing layers. Where sheets parted rather than overlapped, ice developed in a smearing pattern, as if the moving skin was scraping ice onto the water. As I looked through the camera's long lens, tiny dots began to appear all over the surface as minute frost flowers blossomed across the thin film. I looked into the distance to see how far the ice extended.

With a few noticeable pools of water yet to freeze over, golden vapour plumed into the air and evaporated as minke whales made their final retreat north. They must have been able to sense winter was coming and that their breathing holes were freezing over. A flock of Antarctic terns fluttered overhead in a hurry, also heading north to flee their solidifying feeding grounds. Before my very eyes the landscape was finally beginning to transform. It was all happening on one day. It had felt like an eternity waiting for the ice to form and I prayed it would hold. Gazing across the bay to the area to which the emperors would hopefully return, I scanned for icebergs. We'd been extremely unlucky. But in the central part of Atka Bay to the east, a cluster of what looked like twenty or thirty icebergs appeared anchored in position. Despite being a few kilometres away, I couldn't complain. It meant I would have something else to film and focus on that would help illustrate the beauty and majesty of the place that the emperors call home.

The following week was tense and I was desperate for the conditions to remain settled. I downloaded new satellite images as soon as they became available online to see how far the sea ice had extended, and monitored weather forecasts like never before. Even being in such a remote location, weather predictions were incredibly accurate. All I was bothered about were wind strength and cloud cover. Clear days would ensure the temperature remained low and calm conditions guaranteed the sea swell wouldn't break all the ice up again. I continued to visit the edge of the ice shelf to check the situation myself and despite the surface layer of ice rupturing, it wasn't melting or shifting. The remaining patches of open water gradually began to shrink. Looking down on the sea ice, it looked like an image of a city from space, ridges connecting panels of ice like roads connect towns.

I speculated as to where the penguins were and whether they knew their home was being rebuilt. How long would it be before the ice was strong enough to hold one, let alone the entire colony, and would they begin to inspect the ice before it was ready? Whether they'd actually return at all did go through my head on a number of occasions. The emperors had never failed to return to Atka Bay to breed, but I couldn't help but think that maybe this could be the first year they'd choose not to. Looking at satellite imagery, similar if not better-looking breeding sites weren't far around the coast and I worried the penguins would choose those instead. I began to wonder if a huge iceberg out in the mouth of the bay had grounded and blocked their route in. They seemed like crazy ideas but anything was possible. If previous years were anything to go by, their return was expected any day. The anticipation was building. But in the back of my mind, I couldn't get rid of that 'What if . . .?' doubt.

One calm morning, Stefan and I had kitted up and made our way to the ice shelf edge to see if any penguins had started to return. It felt like any other day. Following some heavy snowfall the night before, thick dark clouds began disappearing in the morning sky; a lavender hue lit up the horizon as the low sun climbed gradually higher. As I did first thing every morning, I scoured the horizon with my binoculars. It had become my morning ritual and scanning from right to left, nothing seemed out of place. The same icebergs remained in position and a spire of pink ice reached into the sky, rippling in the hazy air. Icebergs on the horizon appeared taller than ever before, the cool air creating an inverted image, suspended above. I looked back towards the station. Its stilts had trebled in length and despite it being almost eight kilometres away it was as if I was looking back at my house from the bottom of the garden. The atmosphere was eerie and something started to feel different.

In the distance, weird black specks lined the horizon. Against the purity of the white landscape they stood out. Through my binoculars it became clear: the emperors were finally beginning to return. Appearing from over the horizon, they were heading straight in my direction.

Stefan was by my side, but I was desperate to share this unique occasion with Will, too, and he was working back at the station. I grabbed my radio.

'Will. You've got to get down here,' I said. 'It's just magic.'

Before long, Will arrived on his skidoo with Dr Tim and within half an hour the first penguins were arriving onto the area of sea ice below where we stood. Despite snow having carpeted a layer over the fresh ice, a couple of pools of water hadn't quite frozen over. Where tidal movements in between the frozen ocean and the ice shelf had prevented the cold temperatures from taking a firm grip, a strip of loose ice extended around the perimeter at the base of the cliff. I'd read countless accounts of emperors marching enormous distances to reach their breeding grounds, in some cases parading one hundred or more kilometres across the newly frozen ocean and navigating enormous icebergs to arrive at their preferred spot, and for many of the birds in front of me, this was the case, though I couldn't tell exactly how far they'd come. One thing that had made life easier for the returning penguins was that the sea ice was fresh; instead of old pack ice with tall ridges and obstacles for the birds to navigate around, a clean, new layer of ice had frozen across the body of open water. This presented them with an almost pristine flat surface, which made travelling simple and quick for the clumsy birds.

As I glanced over at a calm open area of water surrounded by pure white snow and ice, I wondered why it hadn't frozen. No sooner had I pondered this than it erupted, just as when the orcas had come through a month before. Emperors. Rather than marching, some groups of penguins were *swimming* to their breeding grounds, using each patch of open water to surface and breathe. It seemed much more sensible and energy-efficient to swim rather than walk. It made perfect sense, but I couldn't help but feel that was cheating. Each buoyant bird emerged on the surface of the water from the dark depths under the ice with water rolling off its incredibly repellent feathers. Before arching its back and diving down again, each bird opened its beak wide and gulped a huge lungful of air. I'd never seen a bird out of breath before, but as their sharp mandibles parted I could hear them take huge gulps of air. I sat down and looked into the distance as dark silhouettes continued to roll over the horizon. Each bird had been gorging on up to six kilograms of fish and squid each day during the late summer in preparation for the breeding period. Finally, they had started to return. 'Where have they been?' I asked the guys who sat beside me, in a relieved manner.

Over the following weeks long lines of penguins continued to wind their way across the barren landscape towards their breeding grounds, just a few hundred metres off the edge of the shelf. By this point the area they had congregated on contained thousands of emperors and still more birds arrived. It was beyond my wildest dreams to be witnessing such an iconic natural event, but my frustration had started to build. I'd been confined to the top of the ice cliff for months, elevated above the bay looking down, and I was desperate to get closer. But with still relatively fresh sea ice conditions, venturing onto the frozen ocean was still considered too dangerous. Emperor penguins weigh on average between twenty-three and twenty-five kilograms, but I was seventy kilograms and my camera gear added another twenty. It was far too much of a risk. I had thought that the penguins returning meant my long wait was over, but instead I was delayed further, waiting for the ice to become strong enough to hold my weight. Watching the colony from a distance was torture. During some days the light was perfect, the wind was perfect and the snow on which they were standing seemed perfect, but there was nothing I could do other than watch. A lot of the time I felt like clambering down the cliff and walking across the ice, but in the back of my mind I knew I couldn't. No matter how much I wanted to be in amongst them filming, it wasn't worth the risk.

With only seventeen days of light remaining before the sun disappeared for over two months, the pressure started mounting. Time had flown and I still hadn't managed to get onto the frozen ocean. Will, Stefan and I sat down and recapped all the rope access training we'd completed the year before, refreshing our memories in the event of any accidents like falling through the ice. Will started some extra research and made email contact with scientists in the Arctic who specialised in travelling on new sea ice. We were desperately trying to get any advice we could on what our next move should be.

It had been over a couple of weeks since the birds began returning and even longer since the sea ice had re-formed, but as we had no experience on sea ice we couldn't make the call. We had conference calls with AWI experts in Germany, who continued to advise us to wait. We had to do what we were told by the experts, and that meant being patient, something I wasn't very good at.

I started to panic. What behaviour was I missing? I knew we had time on our side as we would be in Antarctica for almost a year, but I also knew that behavioural events wouldn't last forever. The emperors had already had two weeks to find a partner and mate, and with all the birds having arrived I began to think they may all have already paired up. The thought of missing these key pieces of behaviour was terrifying.

Each day, the sun's time above the horizon became noticeably shorter. Under clear skies, shadows lengthened and weather conditions and temperatures began to substantially deteriorate. Strong winds hitting Neumayer became more frequent and the lengths of the storms began to increase. I found myself filling in my diary each evening with the same phrase: 'Day at the station – storm.'

To help me fill such downtime at the station I'd taken with me a few projects from home that I intended to complete, one of which was to edit a short video I'd filmed back on my honeymoon with Becky. With a hard drive full of random clips I'd filmed while riding bikes, scuba diving, standing at the top of waterfalls and in hot air balloons, I'd brought all my material with me to Antarctica to edit together. It had been over nine months since Becky and I had been on the honeymoon of a lifetime and even though I'd promised her I'd put something together, I hadn't told her of my intentions. It seemed the perfect gift to send her on our first wedding anniversary when July came around, so I began to sieve through the mountain of material each evening. It was the first time I had watched the short clips since recording them and it brought back so many fond memories. Back at home, Becky was less than a month away from giving birth and even though I spoke to her every evening, seeing us together on my computer screen made me realise how much fun we had together and just how much I loved and missed her. I'd not seen her for almost four months and it felt like a taste of normality sitting watching images of us together, the perfect distraction from both an Antarctic gale and the worry of missing key events for our film.

3.

The Beginnings of a Dynasty

For weeks, I watched the colony from over half a kilometre away stuck on top of the ice shelf, frustrated by my lack of access, wondering what the emperors were up to. The penguins had been back for over a month by this point and every day the noise that floated through the air from the colony emphasised that things were happening. Despite the sea ice having re-formed before the penguins returned, there was still a possible risk that it wouldn't be able to support my weight. Obviously, falling through it into the freezing waters would have been a disaster and a risk we couldn't afford to take. I was desperate to get closer but I just couldn't. I found myself getting angry and frustrated at why we'd selected a colony that for such a large chunk of the year was inaccessible. I knew we'd been unlucky with the ice conditions, but for such a unique and unusual film shoot, I asked questions.

'Why have we travelled to the other end of the Earth and sacrificed so much to end up watching the emperors from such a huge distance away?' I asked Will in an annoyed tone, but it hadn't been his decision to use the emperor colony in Atka Bay. All the training, the months of medical assessments and the huge amount of preparation all felt like a waste. I wrote positively phrased emails back to Miles in the UK emphasising how patient I was being and that there was nothing to worry about, yet deep down I felt the situation was becoming desperate. Miles repeatedly replied with, 'You can only do what you can do.' He understood the situation and I'm sure he was as nervous as I was, yet he didn't let on. We'd barely made any progress on the film other than a few shots of distant birds arriving across the new ice and I was petrified I was missing important behaviour that the beginning of our film was relying on.

Becky took the brunt of my anger over the phone, but being just two weeks away from her due date, that was the last thing she wanted to hear. 'So you could have been here with me?' she questioned. In hindsight, it wasn't as bad as I was describing to her, but I'd allowed myself to get worked up and I couldn't bear the thought of failing before I'd even begun. The tension started to build; I'd put everything into this trip and I was determined for it to succeed, but right now it wasn't working.

Even though we had still not been granted authorisation by the German authorities to attempt a trip onto the sea ice to reach the emperor colony, Will, Stefan and I started preparing to make sure we were ready for the day they did. Part of our mandatory kit that travelled everywhere with us as a safety precaution was a large green aluminium box, screwed down at the skidoo end of the wooden Nansen sledge we towed behind us. Its contents were indicated by the large white block capital letters down each side: MOUNTAIN RESCUE. Four aluminium bars, two strips of hardwood and four heavy steel bolts secured it down. We'd used the ropes and harnesses during the summer when filming close to the edge of the ice cliff but fortunately we hadn't had to break into the box since. Also inside it were ice screws, crampons, a rescue sling, an avalanche probe (a pole similar to a tent pole, which was used to feel for crevasses under the snow) and thick blankets wrapped in plastic bags. Everything that you could think of that would help in a rescue situation was to hand. One box contained enough kit to look after all three of us and even though I felt confident we'd never have an emergency in which we'd need any of it, knowing it was there meant I could relax.

With under a fortnight of sunlight left, I felt we had a lot of catching up to do in a very short period of time. The penguins had returned weeks ago to begin their annual winter breeding process and I was extremely worried I'd be missing key behavioural moments before I had even got close to them. Unable to see exactly what they were doing from such a distance, I couldn't tell what stage of the process they were at. The stretch of sea between the bottom of the ice cliff and the sea ice had been the last to freeze, but as temperatures had remained consistently lower than −20, the area appeared to have solidified. Although the newly frozen ice at the bottom of the cliff was hidden under a layer of soft snow, we decided to attempt to get down onto the same level as the penguins. We weren't allowed to venture any distance onto the sea ice but nothing was mentioned about us not being able to abseil down the ice cliff while remaining roped to the top. I wouldn't get any closer to the birds but it was at least a different angle to film from and it was a step in the right direction.

We grabbed a long two-stage aluminium ladder from the garage and ratchet-strapped it to the top of the mountain rescue box. At the edge of the ice cliff, with the penguin colony in the distance, I fed my huge orange snow boots through the leg holes in my harness and pulled it tight. With one end of the rope secured to the safety anchor of two skidoos parked next to each other, Will and Stefan helped lower me down. The cliff of ice nearest the penguins was only about five metres high, compared to the Northern Pier where it fell away at between twenty and thirty metres. In some places where crashing water had eroded the cliff over the warmer summer, a steep ramp rather than a vertical drop offered a way down.

Abseiling down, I lost sight of the guys as my head disappeared below the lip of the ice shelf. Right underneath me enormous blocks of ice looked like they'd been dumped from a truck over the edge, so perfectly rectangular that they looked almost unnatural. The tidal rise and fall that had previously prevented the sea at the base of the cliff from freezing had piled up blocks of ice the size of small tables.

'I'm down,' I shouted up to the others. My boots sank into a few inches of snow, the soles landing on uneven hard ice underneath. I slackened off the rope a little and began clambering over the huge blocks to reach what looked like the flatter surface of the sea ice. The dark mass of the penguin colony ahead of me seemed a lot closer and louder, and even though I hadn't advanced any nearer, it all of a sudden looked within reach.

A pair of adult emperor penguins show off their
astonishing beauty to one another.

Neumayer III research station.

Polarstern, the German icebreaker, offloading cargo
at the edge of the ice shelf.

A small group of emperors drift around the edge
of Atka Bay on an ice floe.

Giant icebergs tower out of the ocean as they drift by.

Waving off the last plane for eight months.

A section of North-Eastern Pier breaks away and floats off into the ocean, our skidoos visible for scale.

Long lines of emperors starting to return across the new sea ice.

Filming returning emperors from the elevated ice shelf as they cross
a crack in the new sea ice. (Photo by Stefan Christmann)

A pair of breeding emperors determine whether they're a match for one another.

With their bond cemented, a pair of emperors rest close together in the golden light.

During a winter blizzard, penguins on the exposed side
of the huddle hunker down.

After a week of storm-force winds the colony of males, each with
an egg on its feet, marches back to where they started.

Having relocated after the storm subsided, the colony
tightly reformed on the sea ice.

A perished egg lies frozen on the ice following a storm.

Will and Stefan lowered the ladder down and kicking its feet into the snow I climbed up it, inserting ice screws into the ice cliff to secure it. Even though I was secured by a rope, I was effectively standing on the sea ice for the first time. It felt solid and all I wanted to do was unclip my harness and head out further. A crack a few feet wide had opened up parallel to the ice cliff, but inside the crack there was ice rather than water. The temperatures were so low it seemed to be solidifying everything. In reality I'd managed to get no more than ten metres closer to the birds but it felt like real progress and was of huge psychological benefit.

Although none of us had any real experience on a newly frozen ocean, having spent the previous few months filming on the ice shelf I'd learnt a lot about the challenges and risks. At the evening meeting with the base commander, Dr Tim, we thought about trying to push for official access onto the sea ice. I felt that using the ladder each day to get down to our lower position could be done safely. We came to a joint agreement that while we waited for formal permission to come from the powers that be back in Germany, I could be lowered down to and film from the bottom of the ice cliff, so long as I was roped back to the safety of the ice shelf. The other condition was that I was the only one down there.

Personally, I didn't have any worries about falling through the ice. It was, after all, managing to hold a few tonnes' worth of emperor penguins without much difficulty. Despite having seen the bay free of ice over the summer, with all the open water now hidden, I'd lost sense of the potential danger. It was as if it wasn't an ocean any more. Out of sight, out of mind. I didn't feel scared at all. All I was concerned about was what the penguins were doing. I knew I only had one chance at capturing the courtship behaviour and every day that I missed, I felt that getting the shots was becoming less and less likely.

Over the following days I roped myself up and climbed down the ladder, edging further and further out onto the sea ice. My rope was fifty metres long and even though I still wasn't officially allowed access onto the sea ice, I worked the system, knowing I was secured to the safety of the skidoos up on the more stable ice shelf.

The morning of 15 April started early. At 6am a heavy knock on my door woke me up. With the sun not due to rise for a good couple of hours, I thought I was dreaming, but it was Will, standing half naked, his eyes still adjusting to the bright lights having been woken up unexpectedly. Unable to get through to the phone in my room, Becky had called Will in an attempt to get hold of me. Embarrassingly, rather than having the cable plugged into the phone in my room, which would have allowed Becky to connect to me directly, I'd had it providing my laptop with an internet connection overnight enabling me to download an episode of the BBC's *Traffic Cops* to watch during the next storm. Her due date wasn't for another four days and I wasn't aware of how quickly things could change. I definitely wasn't expecting a call to tell me Becky had gone into labour. Once Will had handed the phone to me, Becky and I chatted and we both decided it was best to carry on my day as planned. There was nothing I could do anyway, other than wait by my phone anxiously.

Later that morning, I was down on the ice trying to keep my mind off what was happening at home. I looked behind me along the full length of climbing rope, clipped to my harness, which lay across the ice between me and the cliff. I pulled it tight with my waist to get as far out onto the sea ice as I could. With one hand carrying the camera and the other holding onto my baby tripod, which allowed me to film as low to the ice as possible, I got set up.

Next to a large fresh crack that led off into the distance, an enormous grey mottled Weddell seal lay singing in its sleep. This was the one other species that Stefan had seen on his previous visit and he had been confident we'd see some. Weddell seals breed further south than any other mammal and, like the penguins, endure horrendous weather conditions. However, with the help of their teeth, they regularly gnaw the edges of holes in the ice to prevent them from freezing over. Spending their entire year on the ice, there was a real possibility they'd become trapped either under the ice or on top of it. Feeding on fish and marine crustaceans, they were no threat to the penguins. I'd been desperate to get close to one. Its chorus was subtle and unlike anything I'd ever heard before. It sounded electronic, similar to the old modem dial-up method of connecting to the internet. As the sound was designed for travelling underwater rather than through air, it was difficult to hear initially above the ice. With upturned corners to its mouth, the seal appeared to be smiling and with its head held high above the snow, its whiskers twitched. Every now and then its bizarre, seemingly computerised vocalisations would stop, its concentration switching to scratching its thick-fur-covered body with its tiny claws positioned at the end of its flippers. It was gigantic at over three metres in length and I felt small and slightly vulnerable in comparison, but as it had rarely, if ever, seen a human, it was unafraid of my presence.

Behind the seal a small group of emperors that had ventured away from the main colony had been halted in their tracks by the crack. We'd watched them happily toboggan along on their bellies on smooth ice, using their wings for balance and their feet as motors. Coming across damaged ice now seemed to come as a shock to them. Quickly, individuals rose to their feet, their sharp beaks pushing against the snow to lift their heavy bodies back upright. They stepped back away from the crack while they assessed its stability. It was obvious to me they knew it was a potential danger and their reaction and small-scale panic made me laugh. Each bird walked up and down along the fissure looking for a way across as more birds were attracted over, intrigued by the situation. With emperors pacing back and forth just a few metres away, the Weddell seal didn't even flinch.

All of a sudden, one penguin spotted a route across. It leant forward over the gap, and using its sharp claws and powerful feet for grip, it propelled itself over, landing clumsily on its belly. It immediately pulled itself over the crack and away from danger. Without delay, all the other birds followed. Forming a tight queue, they impatiently waited their turn to cross the split. One by one, they leaped across. To me, the crack looked like nothing. It was substantial, but with such a thick layer of snow on top of the ice it seemed stable. Rather than liquid water in between, the extreme cold had formed a platform of ice, which had solidified the whole length. But the penguins clearly thought otherwise and their intelligence surprised me. Whether it was safe or not, they were very cautious around it and I wondered why. During the summer, in areas where the emperors enter the water to feed, either through small holes and cracks in the ice or off the edge of the ice sheet, leopard seals lie in wait, ready to pounce. Maybe this was their reasoning for being so cautious rather than actually worrying they'd fall through the ice?

Before I knew it, all the birds had crossed and their excursion across the ice surrounding the colony continued, leaving only the sleeping seal in front of me. Standing up on the ice I straightened my legs, the camera about knee height on its tripod below me. Across the flat sea ice, tracks left behind by the travelling emperors reflected the light from the golden sun. Footprints, wing tip imprints, claw marks and the long winding indentations from the penguins' round bellies littered the landscape.

With all the commotion and distraction I'd not had time to feel the temperature in which I was working, but all of a sudden I realised how cold I was. There were clear skies and no wind; we'd left the station with the thermometer reading −35 degrees Celsius. It was the coldest it had been since arriving and I could feel it. My feet were numb and I had no idea whether my toes were still attached. My hands inside my enormous thick mittens had instinctively clenched into a fist in an attempt to reheat my fingers. With no balaclava so that I could film more easily, the skin on my face was exposed. I walked back towards the cliff where my rope was anchored to the ice to try to get my body to respond and generate some warmth. I felt nothing at all; the cold had penetrated to my bones. Will asked if I was all right. A ten-centimetre icicle swung from side to side from the tip of my nose, while a thick white icy crust coated the hairs on my face and the side of my hat, where moisture produced from my exhalations had solidified. 'No!' I replied. 'How on earth are we going to cope during winter if this is early autumn?' I said to Will.

It was without doubt the most uncomfortable I had ever felt and for the first time in my life I was so cold that I actually felt pain. For some reason I'd envisaged that I'd feel the cold up to a point, and that any colder than that, I wouldn't feel, but I couldn't have been more wrong. The difference of just five degrees was enormous, and the more I thought about how cold I was, the more pain I felt. We'd been away from the station for five hours and it was only early afternoon. I tried everything to warm up: jogging on the spot, star-jumps, vigorously rubbing my legs, arms and head, but nothing worked. Knowing the warm station was just a short skidoo drive away was the only thing that kept me going. It was all about surviving.

All of a sudden my radio crackled into life. I was amazed the rechargeable battery still had enough juice to power it when it had been exposed to such horrendous conditions in an outer pocket of my polar suit. It was Dr Tim back at the station.

'Lindsay, I think you better get back,' he said in a happy yet authoritative tone. Was he referring to Becky? Just two weeks before, I'd knocked on his office door and sat down to tell him that Becky was expecting. I didn't want to tell anyone else, but Dr Tim had to know so that he could be a point of contact for Becky's mum in case I was out of touch on the ice when she went into labour. 'Wow, that is fantastic news,' he had answered when I told him I was going to be a father, as if we weren't all isolated at the other end of the planet.

Will and Stefan helped me and the kit back up onto the top of the cliff and within a couple of minutes I had attached a sledge to my skidoo and was off. Back in December, two weeks before our trip, I had broken the news to Will and Stefan over lunch in the UK. Them being the closest people I'd be working alongside, I felt they ought to know, Will especially, as I knew he'd support me if I needed it. They both knew exactly what was happening.

Travelling alone on the ice was forbidden but the weather, despite being cold, was clear, visibility was as good as it could get and the station was only about seven kilometres away in a straight line in front of me. The journey, which usually took around ten to fifteen minutes, was over within seven and a half, most of which was spent airborne as my speed over drifted patterns in the snow took me off the surface. I abandoned the skidoo outside the front door and ran upstairs, throwing my snow boots and outdoor clothing into the corner of the cloakroom as I went past.

Our IT technician was outside my room waiting with a laptop and a bunch of cables in his hand. He hadn't a clue what was going on but had been instructed by Dr Tim to set up a computer and get me connected via Skype. Despite the internet being capable of video calls, I'd not managed to speak to Becky face to face over my phone. This was my first opportunity to see her since December. Within a couple of minutes a dial tone was sounding. On the other end of the line in the UK, as I waited for a connection, was I about to become a father? I was 15,000 kilometres away for heaven's sake.

Finally connected, a pixelated blurry image appeared on my computer screen. The picture froze and restarted but the audio was clear as day, the sound of a screaming little baby dominating. Literally seconds before, at 4.46pm, Walter Arthur McCrae had been born. I sat back in my chair alone in my room and stared at the screen, trying to comprehend the chaos that was appearing in front of me. All of a sudden Becky's face filled the screen, her mum and cousin accompanying her.

'I wish you were here,' she said to me, shedding tears of joy at the birth of our new baby boy, but also tears of sadness that we weren't together. My own tears began to trickle down my cheeks. I grabbed my small compact camera from the desk and took a quick photograph of the computer screen: Becky's eyes were black with smudged make-up, her hair was damp. But in her arms was the most beautiful baby boy.

The whole experience was hard to get my head around as I was so far removed from what was happening. I think the difficult decision to be apart for Walter's birth was made easier by the fact that we had no idea what to expect. This was a first for us both and we didn't know what I would be missing. Despite it being such an incredible experience, I felt quite far removed from what was happening. I don't know if that was something I did on purpose to protect myself from all the emotions or if it just takes time to build a connection with your newborn. Either way, it was all very surreal. I didn't say anything to Becky but I couldn't really believe I was a father. It was going to take some time to sink in.

That evening at dinner, following a couple of early celebratory drinks at the bar with Will and Stefan, I made my announcement to everybody. I'd become known around the station as a bit of a prankster, constantly making jokes and playing tricks on people. I liked having fun and telling stories to keep morale high. One of my best-known stories concerned a very distant family member discovering Antarctica back in the day but not telling people about it. Telling them I'd become a father that day wasn't going to be straightforward. For almost an hour, I did my best to convince them, Dr Tim smirking at the end of the table knowing, with my reputation as a prankster, the trouble I'd have.

'You've just downloaded that off the internet,' said the electrician as I showed him a photograph.

'Bollocks, McCrae,' came another comment.

After many confused looks and much laughter, it finally sank in and they all congratulated me. It was a Saturday night, which meant nothing to me as I just worked depending on the weather rather than the day of the week, but to the others the weekend was the weekend and this was even more of an excuse to celebrate. 'Mary's Boy Child' by Boney M blared out from the tannoy speakers placed on the ceilings throughout the station. 'A king was born today . . .'

Festive themes had become the norm when celebrating any occasion in Antarctica, the cold, dark snowy scene outside making it seem suitable, but that night, the guys made an extraordinary effort. They made me feel so special and I think they were grateful for an excuse to celebrate. A little nugget of reality had been thrown into our most unusual and extraordinary lifestyle and it reminded us all of home.

At midnight, the team surprised me by leading me up to the balloon room on the top of the station. The meteorologist had filled one of his weather balloons with helium and was preparing to launch it. I wrote Walter's full name on the side together with the date and our location and threw it into the cold night air. Watching the balloon float off into the sky, I still couldn't believe that I'd become a father. The news was going to take some time to sink in; in one way my life had changed forever, but in another, nothing had changed at all.

'Permission granted' were the words coming out of the speakerphone in the office. Finally, after weeks of anxious waiting, we'd been given access onto the sea ice. The powers that be back in Germany had been monitoring satellite radar images and recent temperatures over Atka Bay and finally decided they thought it would be safe enough to venture onto. Having had a few days of rough weather, Dr Tim, Will, Stefan and I sat down to devise a plan, with the aim of creating a route to the penguins and ensuring the sea ice was stable enough to film on.

The easterly wind had blown a few metres of snow against the base of the ice cliff, burying the bottom half of our ladder we'd so carefully installed. In fact there was so much snow that rather than needing to abseil down the cliff of ice, there was now just a step to jump down onto a slope of snow that led all the way down onto the sea ice. Will and Tim roped themselves together with twenty metres of rope between them and set off across the sea ice, a large drill and tape measure to hand on the Pulka sledge being towed behind them. Roping themselves together ensured that if one person fell in, the other could pull them out.

Stefan started to dig out the ladder while I packed up some camera gear to film with as we followed Will's route. As Stefan reached the bottom of the ladder it became clear how much movement there had been between the ice cliff and the sea ice. Even though the landscape looked solid and motionless, the buckling in the aluminium at the base of the ladder suggested otherwise.

Every hundred metres, Will and Tim dug into the snow until they reached ice and then drilled through until saltwater spouted out from underneath. At each point there was consistently thirty centimetres of ice, stable and strong enough to support the weight of a large car. Will filled each hole with loose snow, kicked it down and recorded the location on a GPS. It was a lengthy procedure, but we were making progress at last.

As we made our way closer to the colony it was an exciting moment for us all. After over four months of waiting we were getting closer to the penguins than we'd ever been before and it felt as if we were able to make a proper start on what we had travelled all that way to Antarctica for. Feeling restless and excited, all I wanted to do was head straight in the emperors' direction. However, I had to be patient. Being on new sea ice, not knowing the depth of the ice I was standing on, meant we had to make sure we were safe, and that took time. Although I was incredibly excited, I felt anxious. The colony had been in place for a long time already and, worrying about what I'd missed, I wondered what we would find.

Emperors begin their breeding process as soon as they arrive at their colony and having been so far away, stuck up on the top of the ice cliff, I'd been unable to see in detail what stage of the breeding process the emperors had reached. I literally had no idea what to expect. As we made our final approach, two birds standing separately from the colony caught my eye. Despite their distance from the colony, the low pink sun stretched their shadows so far to the south that it linked them to the main group. They appeared fidgety and restless. One bird circled the other and as it bowed its head to the ground it drew my eyes down to a bulge on the other one's feet. A swelling on a penguin's belly usually suggests the presence of an egg. I blinked a couple of times, trying to make sense of what I was seeing. Not only did it appear that the birds had paired up, courted and mated, they'd started laying eggs! Had I really missed all this behaviour? Three enormous milestones in the penguins' lives, which our film was relying on, had already passed. All that time and all that effort, yet before we'd even started, I felt I'd failed.

My previous excitement turned to despair. How was I going to make a film about the entire life cycle of an emperor penguin without images of the next generation being produced? I turned and looked towards the main colony, only a hundred metres in front of me. Wanting to film courtship behaviour in chronological order, I decided to leave the couple with the egg and to try to find an individual who had not yet paired up. I'd read that emperors usually find a mate within twenty-four hours of reaching the colony and with all of the penguins having returned, I wondered if my chance had slipped away. But I was keen to try anyway.

In my head, working in chronological order would make filming the story easier. I'd know what I had filmed, what I needed to film, and what I needed to film more of. As my chin fell forward onto my chest in dismay, I looked towards my feet. Leaning forward, I transferred my weight against the harness of my camera sledge and started walking towards the main group, leaving the couple with the egg behind me. Within one minute I stopped and looked back; the two birds still seemed as unsettled as before. For some reason I changed my mind; the lure of an emperor with an egg was too strong.

'Let's just give it half an hour to grab some shots,' I said to Stefan.

With the camera set up I knelt down next to it and put my eye to the viewfinder. I tilted the tripod head down from the pink sky onto the faces of the two emperors, each bird facing the other in a moment of calmness. As I steadied the shot, one bird bowed its head. I went with it, watching as the bulge on the other one's belly rose and filled half my frame. As if in slow motion, each feather covering the bulge began to unravel. Lit up like diamonds in the glowing light, the white feathers began to reveal the penguin's egg. With the camera still running I held my breath to keep the image as steady as possible and prepared myself for my first ever glimpse of an emperor penguin egg. What was revealed amazed me. A large circular snowball. The penguins were incubating a block of ice!

With such a strong maternal instinct, using their beaks to push bits of ice onto their feet isn't unheard of in emperor behaviour, and it was something I was desperate to see and document, but to watch such a clear example play out in front of me, with a ball of snow the same size as an egg actually being incubated upon a penguin's feet, was extraordinary. I smiled to myself as a huge sense of relief washed over me as I realised I hadn't missed all the courtship behaviour after all. It was one of those first magical moments that I'd been unable to plan for but had hoped to see, and it made all the effort and uncomfortable time we'd spent outside in increasingly difficult conditions worthwhile.

I wondered if seeing the behaviour so clearly was a filming first. How many more times would I see it demonstrated in amongst the other 10,000 birds? The pair had given us a captivating piece of romance on film before we'd even properly started and it gave me an enormous boost after a few weeks of feeling mentally pretty low. Having been concerned that it was becoming impossible to document everything in the emperors' life cycle, we were back on track and my chronological method of working through the different stages of the penguins' breeding process could begin.

As soon as emperor penguins re-form their colony each autumn, the timer starts ticking. Reproducing on a temporary surface that often melts away during the summer, the colony enters a race against the sun. They must raise their chicks to an age where they can support themselves before the sun generates enough warmth to melt and break up the ice. Due to this, emperors waste no time in starting the breeding process each year. Rather than searching through the whole colony for their previous year's partner, the emperors find a new mate every year.

Finally in position at the colony with my camera, I fixed my lens onto a single bird. Having spent the summer at sea feeding, he or she looked magnificent: fat, tall and incredibly vibrant. The golden neck patches glowed. Black feathers covering its head and back reflected a dark metallic blue sheen. Each feather led to an immaculate sharp point. Male and female emperor penguins are known to look incredibly alike and it made me determined to find a difference so that I could distinguish which was which, to prove the theory wrong. Though the harder I looked, the more I struggled. Each bird seemed identical.

My bird entered the arena and I zoomed in on the top half of its body and started following it. Within a few feet it stopped and threw its head forward with such vigour that its long beak bounced off its chest, springing up and down before coming to rest. It leant forwards until its beak nearly made contact with the ice, seemingly defying gravity as it remained balanced on its two flat feet. Parting its beak slightly, it inhaled as deeply as possible and following a short pause rapidly straightened its body, exhaling all the air and producing a long, almost unnatural, trumpet-like call. Set alight by the low sun behind, its warm breath resembled fire as it reacted with the cold Antarctic air. Slowly, as the penguin stood tall, it rose upwards revealing its ankles and short legs beneath the pristine white belly feathers. Pointing its beak skywards, the whole body elongated, increasing its height by up to a third. It froze, standing motionless amongst the chaos of thousands of other penguins, listening.

I zoomed my camera closer in towards its eyes. They were wandering, the only movement in the penguin's body. For almost a minute it resembled a statue, standing tall, every muscle in its body taut, holding its head high above the rest of the colony. It didn't even flinch. It was waiting for a response from another penguin prospecting for a compatible partner. Nothing. As quickly as the bird had come to a stop, it relaxed, retracted its neck and walked on. Within ten seconds it stopped and repeated the process. Still nothing. Most of the birds, having been in position for a few weeks, were already paired up, but with the odd bird still searching amongst the colony, I felt there was still a chance to capture the entire courtship process.

I looked across the sea of birds in front of me, my eyes jumping from one bird to another, picking out birds that were strutting through the tightly packed group. There were clearly still a few birds that hadn't found a partner and the intensity of the call of these single penguins suggested the same. It was deafening. How on earth were the emperors able to distinguish differences between individual calls, let alone isolate and focus in on just one of those calls?

I continued to track my penguin with the camera through several calling attempts, but each one met with no luck. The bird popped out from the main group and walked towards me to try its luck along the edge of the colony. I tilted the camera downwards and followed its reptilian feet as they stepped across the ice. Dark-grey and scaled, a long sharp black claw protruded from each toe. The feet stopped. With claws crunching into the snow one by one, they lifted the bird as it called. My right eye stared down the viewfinder in anticipation. A pair of feet filled the right of my picture, empty ice on the left. I concentrated and waited, just like the penguin. As I went to stop the camera, a second pair of lizard-like feet stepped into my picture. A reaction. Had a partner been found?

The birds inclined towards each other, their beaks crossing in the centre as they leant forwards to call. Stretching tall to show off their golden shoulder patches, they pointed their beaks skywards to reveal the stunning pink and blue under-side of their bills. Both birds flung their heads backwards at lightning speed, before bringing their beaks back to the centre slowly after making a circling motion with their necks. Their sharp bills rose again, crossing and resting on each other in a touching moment of togetherness. Everything one bird did, the other mimicked. The synchronised ritual was taking place without a moment's hesitation, and I began to wonder if the whole procedure was predetermined, innate behaviour.

Emperor penguins mimic each other in a graceful sequence known as a dance to strengthen the bond between pairs. If one preens its tail, the other copies. If one calls, the other calls. Even when walking across the ice, both penguins sway from side to side with remarkably matched timing, each bird's foot falling onto the snow in the same tempo. Although I couldn't tell the difference between individuals, clearly each penguin knew what they were looking for and watch-ing them pair up and choose their mates felt like a real privilege.

Having finally started to make progress, when bad weather hit and forced a hiatus in filming I didn't feel the same level of pressure. It was a relief to have something in the can and helped me mentally cope with the periods stuck inside. Although frus-trating, periods indoors gave Will and me the opportunity to review footage. Finally having something to send back to Miles to show him what we'd managed to capture felt fantastic. The satellite internet connection at Neumayer wasn't the fastest but we could at least get low-resolution footage back to Bristol for them to review.

This was the first time a film crew overwintering in Antarctica could send footage back over the internet. Previously, crews such as the one that overwintered for *Planet Earth* shot all their material on film cameras and were unable to review anything, let alone send any of it back. They had to wait until the film had not only been returned but developed to see what they had captured. A terrifying prospect. We could watch our footage back and be sure the cameras were working as expected the second I'd stopped recording. It also gave us the chance to edit together short sequences, highlighting shots we were missing, or identifying what would make each sequence better. We went through everything in detail together ensuring we'd managed to cover all the behaviour we wanted.

Having such good lines of communication made me feel slightly embarrassed. Not that many years previously, people had to put up with much tougher conditions and communication would have been minimal; in fact, short monthly faxes were the order of the day. Now here we were sending footage back just hours and days after filming it. Previously, overwinterers wouldn't have even had the opportunity to talk to family and friends back home. It all felt like such a massive privilege.

Even though we had finally made it to the colony and started filming, we'd only just managed to catch the tail end of the courtship display and knew if we didn't ensure that we wrapped that sequence up within a couple of days, we'd have missed it for good. The next time the penguins were going to demonstrate this behaviour would be in twelve months' time and I definitely wasn't going to be here then! Just as we were finishing reviewing some footage, the meteorologist knocked on our open door.

'Lindsay, I have Walter's data,' he said, handing me a piece of A4 paper. Printed on it were two graphs entitled 'Radiosounding Neumayer III Station'. I thanked him and looked over the information. The balloon I had released on the day Walter was born had sent back information. Reaching an altitude of thirty-one kilometres, it had burst under temperatures of −60 degrees Celsius at wind speeds reaching ninety miles per hour. What a fabulous souvenir! I immediately sent Becky a photo and stored it safely in my room on my bookshelf between two pages of a penguin book I had brought with me.

Even during rough weather I'd started leaving camera kit down on the frozen sea so I didn't have to exert much-needed energy on packing it up each evening, tying it down onto the sledge and pulling it all the way back to the station. It wasn't going to get stolen, I knew that for sure! Having established a safe route off the shelf down onto the sea ice, Will, Stefan and I had marked a track with eight red flags, four on each side. Still, the large step down was too much of a drop to risk trying to cross with a skidoo so we relied on passing kit down by hand, packing it into two Pulka sledges and walking the one kilometre in a south-westerly direction across the ice to reach the penguin colony. To make getting up and down easier, a fifty-metre safety rope with knots tied in it at intervals for grip was draped over the step down between the flags, secured at the top by two ice screws, buried in the surface of the solid blue ice.

In the beginning, I considered pulling a Pulka sledge full of camera gear towards an emperor penguin colony to be an honour. In the film-making world I'd look heroic, but the novelty rapidly wore off. Especially in temperatures that sank to −40, the physical exertion soon took its toll. My tall, wide, heavy snow boots meant my gait had to be wider than normal to stop my feet from tripping over each other. The shoulder straps of the sledge dug deep into my collarbones under the weight of all my camera gear. Within a few hundred yards, despite the low temperatures, my thermal undergarments were drenched with sweat. The effort involved in just moving with so much clothing on forced my heart rate to increase. Breathing heavily though a saturated balaclava, I reached up and pulled it down, exposing my nose and mouth. As if I'd been submerged underwater, I gasped, filling my lungs with air. I'd always classed myself as relatively fit but these extremes were starting to push me.

On reaching the colony I removed my hat and pulled the top half of my red polar suit away from my chest, the arms hanging off the elastic straps over my shoulders. Within seconds, I felt cool as the heat was sucked out of my clothing. Although it was a relief to feel cool, any accumulated moisture on the fabric exposed to the air froze instantly, creating a layer of silver frost. For the rest of the day my clothes were wet and trying to work while my body felt so uncomfortable was not ideal.

It was becoming a constant battle. Every day that I was forced to make the journey pulling heavy kit to get to the birds in such extreme conditions, I lost more motivation. Before even getting the camera out of the box I felt as if I had already done a full day's work. All the filming plans we'd made back in the UK in meetings and during training exercises were based around the sea ice remaining in place for our whole stay, and with the conditions having changed, we were having to adapt.

Access down onto the sea ice was usually relatively straight-forward; two ramps of snow, marked at points where the ice cliff was at its lowest relative to the sea ice below, created a safe and easy place for skidoos to transit across the divide. As the prevailing winds blow from the east, loose snow blown across the sea ice over time builds up against the cliff, which in turn forms a slope. This process had already begun, as demon-strated when we nearly lost our ladder, but the accumulation of snow hadn't quite completed the ramp. With access to the colony proving so challenging, the sooner I could get a skidoo onto the sea ice, the better.

During one afternoon of tricky conditions that prevented us from filming, Stefan and I decided to jump in a snowcat and re-groom the seven-kilometre track to where we were accessing the sea ice. Snowcats, or Bullys, are the same monstrous pieces of machinery that are used to flatten pistes across ski resorts and were the main mode of transport around the station. They are immensely powerful and their large black blade at the front knocks away any obstacle made of ice or snow. Drifting snow blown across the ice can quickly form lumps and ripples on the surface. The shapes left on the ground were known scientif-ically as sastrugi, and despite forming beautiful natural patterns, driving a skidoo over them was very challenging and slowed us down. Re-grooming the track took time: the Bully's top speed with the blade engaged was about my jogging pace, but it was necessary as travelling by skidoo across a groomed track could be up to forty minutes faster than having to negotiate mounds of snow. It also gave us something to do during downtime. Sitting inside could get depressing and it was good to find things to occupy us.

As we approached the last hundred metres with the blade overflowing with loose snow that we had ploughed, we had an idea. Rather than piling up all the excess snow at the side of the track, why couldn't we use it to finish off our ramp? We edged the Bully and its load of snow to the end of the track, grabbed a couple of shovels from the back of the cab, jumped out and set to work throwing heavy spadefuls of snow over the step. As darkness fell, the wind still drifting snow across the ice, we bridged the gap between ramp and ice shelf. I sat at the top and slid down it a couple of times. It was steep but nothing a skidoo couldn't handle. Having advanced it as much as possible, we left it in the hands of the wind and cold temperatures to improve the slope and solidify it, making it safe for us to drive down. I crossed my fingers as we drove home. No more pulling sledges, I hoped.

As filming progressed around the ever-deteriorating weather, it started to dawn on me how little time I actually had to get each sequence. Even though I would be there for almost a year, seemingly a gift of endless time in which to produce a film, the weather was starting to make each part of the penguins' breeding process perilously short. The summer had surprised me with its endless sun, calm days and somewhat comfortable temperatures, but as autumn quickly surrounded me, the change was astonishing. It certainly wasn't anything that would disrupt the penguins too much, but days where the wind did blow proved too much to film in, the wind shaking my lens too much to get a steady shot. Three or four days straight stuck back at the station became common as large weather systems blew in and hung over Neumayer for extended periods.

To make the most of good days where we could get down to the colony, we left early before the sun rose, to maximise our filming time, but even this began to fight against us. As we approached the middle of May the sun hovered above the horizon for less and less time each day, until on Saturday 20 May it peeked above the skyline for one last time. Winter had officially arrived. For the following two months the sun wasn't due to make an appearance again, leaving the birds and the station in the darkest, coldest and wildest place on Earth.

Known as 'polar night', it is the period of time when the centre of the sun doesn't rise above the horizon. Occurring in the Arctic and Antarctic circles, 66.5 degrees latitude either side are the sun's limits; any further north or south, polar night resides. Neumayer, located at 70 degrees south, sits well below that line and annually experiences sixty-two days where the sun is below the horizon. For the emperors, breeding in the Antarctic winter, losing the sun just after they'd paired up meant they would have to perform nearly every key piece of behaviour to start their next generation during the harshest, darkest times, with mating, egg-laying, incubation and possibly hatching all occurring in darkness.

The loss of the sun had massive implications for me, too. The camera I'd been using since arriving was not sensitive enough to film under the low-light conditions, so I put together a specialist camera that would hopefully enable me to continue filming even through the darkest days. Most of the modern filming equipment we'd travelled down to Antarctica with was being operated in such conditions for the first time. Technology had advanced so much over recent years that cameras were much more light-sensitive, required less power and recorded for longer than ever before. It was exciting yet tense having such expensive pieces of equipment at my disposal. With no certainty they would actually perform in the conditions, I felt anxious.

During the summer of 2016 I had travelled to a facility in Southampton to test certain pieces of kit in a freezer room where the temperatures had been artificially lowered to −40 degrees Celsius, in an attempt to replicate the conditions in which I'd be working in Antarctica. Assistant producer Theo, our kit expert Gordon and I lined up six different cameras on tripods. We weren't testing the quality of the image they were capable of recording; first of all we wanted to see if they could remain switched on in such extreme temperatures! It was my first time exposed to such low temperatures but it was no comparison to what I would experience in Antarctica. With wind chill and the added challenges of day-to-day life on the ice, nothing could have prepared me, but it gave me an idea of what would work and helped us make final choices about which kit to take. For a bit of fun I took a small plastic model of an emperor and her chick inside the freezer and placed it on the windowsill in front of the line of cameras, to give me something to point the cameras at. I was kitted out with a woolly hat, thick gloves and an old blue ski suit.

With each camera on its tripod, we switched them on and left them to it. Every thirty minutes we re-entered to see how they were coping. Immediately, two cameras failed and one slowed up. I went back in to reassess the remaining three, quickly opening and closing the heavy door behind me. Not only were we testing the cameras, the heavy tripods were also being monitored as these were also vital pieces of kit to keep my long lens steady against the wind. I wrapped my glove around the tripod handle and went to tilt the camera in the direction of my plastic penguin. It was jammed. The fluid in the head, which enabled smooth and precise panning and tilting movements, had frozen solid. I had thought the tripod heads would be the last of my worries but it quickly dawned on me just what problems such extreme temperatures would present.

Gordon had also brought with him a few lengths of Arctic-grade cable for us to test. It was royal blue and extremely thick. Should battery problems arise in the cold, this cable would bridge the gap from an external battery to my camera. In very low temperatures experience had shown me that normal plastic- or rubber-cased cable became so brittle that it just snapped in two and I had no doubt that we'd have this problem in Antarctica. Inside the freezer room, Gordon did his utmost to break the Arctic-grade cable, twisting it, folding it, even standing on one end and pulling on the other. Despite turning stiff, it didn't break. He took it outside into the warmth and closed the door behind him. The vibrations knocked my plastic penguin off its ledge onto the metal floor, where it shattered into hundreds of pieces! The cold had claimed its first victim and I hoped it wasn't an omen for what was to come.

My new low-light camera that I was now using was much bigger and heavier than the one I'd got used to, but it was a necessity for capturing the penguins' big moments under the dark Antarctic sky. With much bigger lit-up buttons it was easier to operate but I didn't have a way of transporting it. In the workshop one afternoon I cut and screwed together a giant wooden box, lined it with foam and tried to seal the lid as best I could. I unscrewed the unique bracket I'd built on the back of my skidoo, which had held down my previous camera box, and cut another frame to accommodate my new box. With the sun having vanished we attached our sledges and began the drive down the newly groomed track to the sea ice. Already the previous day's poor weather had taken away its fresh look; lines of sastrugi cut across the track at ninety degrees. But the snow was soft and the skidoos powered through it, delivering us to the top of the ramp between our flags in no time.

'Shall I do it?' I said to the others. I stood up, peering over the windshield while holding onto the handlebars. I straightened my skidoo and sledge into a line and edged forward. I stopped just at the lip before the incline fell away onto the sea ice, my sledge hitting the back of the tow hitch with a bang as it caught up. I looked down and thought that, even if I ended up rolling the skidoo onto its side but in the process got it onto the sea ice, it would be better than having to drag a Pulka sledge all the way over to the penguins again. I pressed my thumb against the accelerator and went for it, the front of the skis hanging over the edge of the lip before tilting down. My sledge followed, holding its line remarkably well down the steep slope. I accelerated, forcing the route I wanted the skidoo to take, and before I knew it I was down.

'Easy!' I shouted to the others. Stefan first then Will followed my line. Getting skidoos onto the sea ice for the first time was a milestone and we all high-fived.

'Let's get over to those penguins!' Will ordered.

Surveying the colony, as I did every day when I first arrived, I looked to see what the birds were up to. It was quiet; only a few birds still patrolled through the group stopping and calling to attract a mate. The likelihood of them now finding a mate and being successful was very slim and I felt sorry for the ones left behind. I bent my legs and knelt down, my eyes head height with the emperors. I glanced at each bird that stood in front of me and moved my focus from one to the next. All of a sudden my vision fell onto the black sheen of what I thought was a penguin's back. Although I had thought I was looking at the back of a bird I suddenly realised it was looking right at me. I blinked and ran my eyes down its body; a few white specks that looked like snow stuck to its coat. As my gaze arrived at its feet, it became clear they were also pointing in my direction. What on earth . . .? I stood up, and at the same time it opened its wings and stretched. An all-black emperor! Jet-black plumage from head to toe, the only bit of normality being a hint of yellow on its neck.

With a colony of nearly 10,000 birds to watch, I had quietly expected to see something out of the ordinary, but this was beyond my imagination. This was rare and, as I found out later, new to science! I grabbed the camera and started collecting some shots, zooming in on its white toenails. Tilting the camera up the white specks that I had initially thought was snow it became clear that they were white feathers, not many, but as pure white as the feathers on the neighbouring penguin's belly. Despite looking completely different, to me, it didn't stand out that much amongst the other emperors. It just looked like a penguin looking in the opposite direction; I could so easily have missed it. For the first time there was a bird I could identify amongst all the others and I wondered how many more times I'd see it.

We gave him or her a name, Nelly, and I watched it walk off into the main group. The other penguins parted and panicked as it approached, moving out of its way as if it presented a danger. They could clearly see a difference in its appearance and it became clear to me why it was still searching for a mate: no other penguin wanted anything to do with it. Nelly waddled deeper into the group as penguins split apart making way for it. I felt sorry for the bird and urged it back to sea to feed. There was no point in staying for the winter with no purpose, and if it did leave, I wondered if I'd see it again.

Having completed the sequences of finding a mate and courtship, mating was next on the agenda. For a bird so well designed for hunting in the water, just walking on the ice was a clumsy affair. I had a feeling that for these round-bodied individuals, mating was going to be no different. Unlike the courtship ritual, which can continue for numerous days, mating would only last for a short time so the pressure was on. I had seen birds mating at the opposite side of the colony but I had always been either too far away to capture it or my view had been obscured by other birds. It was going to be tough being in the right place at the right time and I started looking for signs that mating was imminent. Basically, I couldn't detect any visible signs, not really. They had to be lying down to mate and the male would try to push the female to the ground, but she would only allow that when she was ready. I just had to hope that I got lucky.

I wanted to capture the whole sequence of events; the tricky bit would be getting the birds falling to the ground just before mating. Kneeling next to the colony, I picked out two birds in front of me. The male had been trying continuously to pull the female to the ground, wrapping his neck around hers and putting his full weight on to her. I took a gamble and thought maybe this couple were close to mating. Next thing I knew, the female opened her wings and leant forwards, lowering herself gently onto the ice, inviting him to mate. Lifting her tail was her sign to the male that she was ready. He seemed surprised; I wondered if he'd come to the conclusion that today wasn't his lucky day.

It was very difficult for the male to climb on top of the female and it was almost comical watching him try. Hurrying into action, he used his beak and feet to climb onto her back. Once he had mounted her, though, it was over in seconds, and the male soon lost his balance, rolling off onto his back in an ungainly fashion. The female looked to her side to see her partner flapping upside down and, seemingly embarrassed, gave him a poke with her beak. The next generation of emperors was on its way.

That evening, Becky's mum messaged asking to speak to me. I phoned her straight back. 'There is nothing to worry about,' she said calmly. She told me that Walter had been admitted to hospital at less than a week old. Every day since his birth Becky had been sending me photographs and I was starting to feel a connection. From photographs of me at that age there were a lot of similarities and I already felt I knew him so well. But my main concern was obviously for Becky. I had no concept of what she was going through in trying to bring Walter up single-handedly. It had only been five days but the shock of being a first-time mum was hitting, though she wouldn't let on. She knew I was finding it difficult being unable to help and she didn't want to make it worse.

Walter ended up being fine, just needing some help getting food down, but with all the focus having been on me and what would happen if I became ill and was unable to get help, I suddenly realised the reality of my family's welfare back home and how I couldn't help them if they got into difficulties. With Becky taking the opportunity to get some rest I was unable to talk to her, having to go through her mum to see how she was. As when Walter was born, I was encouraged to continue to go out and film to distract my mind. I knew deep down Becky was doing a fabulous job and being so close to her mum and dad, Walter was getting the best possible start to his life.

Finally, Becky sent me a message accompanied by a photo of Walter lying asleep next to her in the hospital. He looked well; he had a tiny white woolly hat on and, tucked tightly beside him, a small cuddly grey penguin chick.

'The past few days have been so crazy I forgot to tell you,' Becky's message read. At the bottom of my case before I'd left she'd smuggled in a small gift, enclosed in penguin wrapping paper, of course. I'd found it when I'd unpacked, placed it on top of my wardrobe and forgotten about it. She messaged again, unable to speak aloud on the hospital ward. 'Open it,' she said. So I did. My very own penguin cuddly toy, identical to the one Walter was cuddling. 'Take it everywhere with you, and remember, we'll be here when you get back.'

4.

The Next Generation is on its Way

In nature, egg-laying is a very private affair not normally witnessed by humans. It usually occurs in a dark hole in a tree, or deep in a hedgerow enclosed by foliage. As it's normally such a private and rarely seen moment, it was the behaviour I was most desperate to capture. I'd scoured the internet trying to find out any information I could about the emperors' egg-laying that would give me an indication of what to look out for.

My only experience of seeing a bird laying an egg had been with a female kestrel back at home in the Lake District. For years she'd nested in the gable end of an old barn, the inside piled high with bales of hay. Over one winter I removed the hay bales to expose her nest chamber and replaced them with a small wooden box I'd built for her to nest in, with a small hole in the back so I could watch her. To the kestrel, the appearance of her nest site hadn't changed, but I was now able to see into her private world. For over a week I watched her from centimetres away, arriving at the barn early in the morning and entering the gable end from where I could see into her box. Every forty-eight hours she laid an egg until she had a full clutch of six. Before laying, she'd fluff up her feathers, stand tall and pant. She'd look uncomfortable and get increasingly restless before suddenly lowering her tail and releasing an egg. It would take her a good hour of calmness and silence sitting on top of her precious brood to recover; the strain on her body was obvious. A kestrel couldn't live further away from or behave in many ways more differently to an emperor penguin, but it was a bird laying an egg and it was an experience I was relying on to help me capture my very own ground-breaking images.

Towards the end of May the colony had fallen very quiet and become relatively inactive. Pairs of penguins stood tightly together having not left each other's side for the previous month since pairing up. The bond between the males and females was one of the strongest I'd ever seen in nature and the affection they showed towards one another was adorable. Rubbing their heads against one another with their eyes closed, pressing their chests hard together, even lying down with a wing resting on their partner's back, they clearly cared for each other.

I'd logged a date in my diary from filming the mating process to remind me of the earliest day eggs would start to appear in the colony. It was only a few weeks but gave me something to aim for. Having missed the majority of birds mating due to not being able to access the colony on the newly formed sea ice, the date I had was extremely rough, but it gave me an idea.

The sun hadn't risen for seven days and we were well into polar night. Having had no experience with twenty-four-hour darkness I wasn't sure what to expect when the sun disappeared for the last time. A lot of people had asked me before I travelled, 'What will you do when it's dark all the time?' Like with most questions I was asked, I simply didn't know until I experienced it. I had been expecting to be under constant stars for at least one month unable to film anything as low light levels would prevent me from doing so, but to my surprise it was completely different. Although overcast days remained dull, when the sky was clear and cloudless, the atmosphere became magical. Over midday we would experience a kind of twilight. Looking to the north when the sun came closest to the horizon, the sky would light up a fiery orange, fading to a peachy hue before being taken over by the dark blue of space. To the south, a band of lavender sat just above the horizon and with nothing impeding my view of the horizon I could see the shadow formed by the Earth in the sky. Although the conditions were brutal, the landscape still offered so much beauty. The specialist camera we'd taken down was worth its weight in gold and even on those overcast days it was still allowing us four to five hours' worth of filming time. The improvements in technology since the last overwinter filming trip to Antarctica gave us the opportunity to film like never before.

A small storm had come over Neumayer and I'd been stuck inside for three days. As during most rough days, I'd easily been able to fill my time sorting, cleaning and repairing kit, and building items that would make my life easier outside. During the few weeks I'd had filming the courtship and mating process in the colony, I'd learnt a great deal about how to make the most of every filming opportunity and it was making the effort I was putting in worthwhile. Each day I selected a pair of birds to try to work with for as long as possible. This enabled me to capture all events in true chronological order and stopped me from swinging my camera around aimlessly, picking up small bits of behaviour from various couples.

Sometimes, for hours on end, I'd have captured the intimate details between a pair leading up to a moment such as mating, and just as the sequence was near completion, with the female accepting the male's mating invitation, my viewfinder would turn pure white with the chest of an inquisitive penguin that had walked over to me and stopped right in front of the camera, blocking my view. As lovely as it was to make new friends each day, it quickly became incredibly annoying and frustrating. Standing up, lifting the camera to one side, sitting down again and re-levelling the tripod head, the whole time dressed like the Michelin Man with a balaclava across my mouth reducing my air intake, proved a huge effort. When I did eventually try to refocus the lens on the couple I'd spent so long watching, either they'd have vanished or they'd be surrounded by other penguins blocking my view.

Singling out a pair of penguins in 10,000 birds was tough. All that time spent concentrating on filming with nothing to show for it started to get to me and I needed to work something out. I racked my brains as to what kit would help me and came up with a novel solution. In the workshop above the garage back at the station I asked the mechanic if he had any spare strips of plastic sledge skids. The white plastic material used on the skis of our sledges was something that needed repairing regularly and knowing that Neumayer stocked enough supplies to fix practically anything, I was sure there would be some spare lengths lying about. The mechanic presented me with a brand-new metre-long strip. I chopped it into two equal lengths and heating one end of each length with a blowtorch I upturned the two tips, making them look like a pair of skis.

With the station's endless cache of wood and a workshop kitted out with every tool imaginable, I set to work putting together a square wooden frame and screwing the plastic strips to the underside, countersinking the small screws that held it in place so they didn't create friction on the ice. I placed each of the three heavy tripod feet at equal distances, drilled holes in the wooden frame and secured down short straps to lock the tripod feet in place. Finally, I screwed one of the two tiny tripod bowls onto the low wooden frame, which would give me the low-angle shots at the same height as the penguins' feet that I was so fond of. A length of rope tied to the front to pull it along and I was finished. The idea was that in the space of just a few seconds I could slide my whole set-up a few feet to one side, bypassing the friendly yet annoying penguins that just wanted company, enabling me to continue filming. With egg-laying imminent, it was a piece of kit I envisaged relying on a great deal. I certainly didn't think I would get two opportunities at filming a female laying an egg, so I needed all the help I could get.

As the storm released its grip one evening, I decided to take a trip down to the birds. I'd missed four days of activity right across the date I had in my diary when I expected egg-laying to commence, so was keen to see if there had been any developments in the colony, and even though it was almost dark, I'd missed being with them. Being kept inside during storms was incredibly frustrating and I was constantly worrying about what I was missing. If there was any opportunity to get down to the birds, I took it.

The prevailing winds had pushed the birds tight up against a tall portion of the ice cliff that towered above them and there seemed to be a lot of activity. The emperors could sense the winds easing and were beginning to spread out across the ice. I looked closely amongst the individuals and noticed some birds stood alone, unlike during the previous few weeks when they'd not left their partners' sides. These lone birds were hunched down and on their bellies, just above their feet, I could see bulges. Eggs! The timing was impeccable, almost to the day I'd predicted. I could see at least twelve birds, shuffling forward, looking cumbersome, with an egg balanced on their feet. Leaving the colony to get back to the station for dinner, plans ran through my head. I was so keen to capture this rare moment and whatever the weather, I wanted to be prepared to give myself the best possible chance.

Scouring books and websites, I tried to find any more information that would help me spot the key moment. From the earliest stages of planning my year filming the emperors, their story and the order of events of their breeding cycle had been running through my head, but to be honest it was all guesswork. Although emperors have been studied in the wild for a number of years, there is still so much that is unknown. I was being forced to learn as I filmed, which at times proved exciting and rewarding but was also frustrating, as it was easy to miss things. I read countless articles detailing the moments that followed egg-laying. The females would head off to sea to feed, but before doing so, needed to transfer their precious eggs over to their mates. But how long after mating would it happen? The amount of key behaviour I needed to film over quite a short period of time began to alarm me. Together with increasingly difficult temperatures and reducing hours of light, everything seemed to be stacking up against me.

The next morning, with the wind blowing just hard enough to unsettle a little snow on the ground, we uncovered the skidoos and headed down to the colony. Cloud coated the landscape, reducing the contrast on the ground to zero. As the sky blended seamlessly into the snow in front of me, the only sight that reminded me I wasn't hovering were the shadows created by my skidoo's headlight. With the prospect of recording such a key piece of behaviour, Will had put work on his computer to one side for the day and accompanied us. I liked it when Will came down to the birds with us as even though his priority involved working back at the station, I felt guilty that he was spending so much time inside sifting through the hours of material I was producing. Liaising with Miles back in Bristol and penguin scientists in the USA, Will was in charge of everything other than actual camerawork, so the only thing I had to concentrate on was obtaining images.

With the clouds getting darker and light snowfall starting we knew we'd only manage a couple of hours' filming, so we hurried over to the birds just before lunchtime. It wasn't too cold, around −25, so rather than being tightly packed together the birds were spread out across a huge area of the sea ice. I loaded my camera sledge and started patrolling the outer edge looking for any birds that appeared to be behaving differently. More penguins appeared to be standing motionless and alone with bulges on their feet. I knew egg-laying was well under way and even though there were an enormous number of birds that hadn't laid, I felt massive pressure.

All of a sudden my eyes were drawn to a pair of emperors, about fifty metres in front of me. Nothing seemed out of place and I wasn't sure why I'd been attracted to this particular couple. I couldn't put my finger on it, but something made them stand out. With Will in the distance to one side also look-ing out for signs of egg-laying females, I lowered myself down onto my knees and turned on the camera. As usual, reducing my height to that of an emperor immediately provoked interest and attracted a bird over to me. Maybe once I was the same size they felt safer to approach me to within touching distance. It stopped directly in front of the lens. With no effort I slid the camera to one side, shuffled up next to it and resumed filming my original couple. Quietly, I joked with the penguin, 'Not so clever any more are you?'

As I zoomed in slowly, I focussed on the feet of both birds. Their heads disappeared out of the top of the picture. Before I could steady the camera, and without any warning whatsoever, an egg appeared between the female's feet! Within the space of thirty seconds from having first locked onto the couple, the female had laid her egg. My camera hadn't recorded any of it. I couldn't believe what I'd just seen. I'd missed it. Leaning back, I stared at the new parents, their large egg, as white as the snow, nestled perfectly on the female's scaly feet. Taking a deep breath, I tried to calm myself down. I felt angry and disappointed. My instinct had told me exactly what was about to happen, yet a small lapse in concentration and a bit of bad luck had lost me my dream shot. Despite having a huge number of birds still to lay, I couldn't help but think that was to be my only chance.

That evening, I felt depressed. I felt as if I'd failed. I replayed the whole sequence of events in my head, praying another opportunity would arise. Not having any footage to analyse, I still had no clues of what to look out for. What did surprise me was the speed of the affair. It all happened much faster than I had anticipated and I knew if I was to succeed I'd have to be ready.

More wind. Yet again, a system of low pressure descended over Atka Bay, forcing us to abandon filming until it passed. Being forced to spend days inside the station when the weather was bad, I knew I had to maximise the periods of good weather. The bad weather days became natural breaks in filming and time for us to rest. Periods outside sapped me of energy so quickly that I needed time inside to recuperate. Settled periods had become shorter and shorter with at least three or four days out of every week too dangerous to venture outside in. It wasn't any of the vicious stuff an Antarctic winter is famed for, which would have affected the penguins, but with the reduced hours of daylight and severe temperatures it made sense to stay safe inside the protection the base offered when the conditions weren't good enough for filming. The reduced visibility that the weather brought meant our cameras couldn't see anything anyway. Bad weather didn't affect Will's work, editing together short selects for me to view and take note of what behaviour I needed to focus on. I'd offer him help to give him a break but to prevent complicating his way of doing things, he always turned it down and marched on.

Every day that I wasn't able to film, trying to exercise was one of the first things I thought of doing to fill time. The process of travelling outside and filming on the ice was physical and extremely repetitive. Sitting for long periods in the same position and carting round heavy kit in extreme conditions wasn't doing my body any favours. Back home I'd always enjoyed exercise outside, especially riding my bike through the quiet lanes of the Lakes and over the mountain passes, and it was something I was missing greatly. It was mid-spring back home in the northern hemisphere, and I thought about the lengthening days, imagined feeling the warmth of the sun against my legs and listening to birds singing as I rode around Coniston Water, a short but beautiful ride straight out of my back door.

Almost every year since turning eighteen I'd ridden the Fred Whitton Challenge, a 112-mile bike ride around the Lake District that attracts over 2,500 road cyclists each May and raises terrific sums of money for local charities. With the event growing in popularity, getting a place had become harder year on year and entering it had become the norm for me every January. Even in Antarctica, I made sure I gave myself a chance of getting a place, submitting my entry online with the aim that, if successful, I would defer my place to the following year's ride when I was home.

In a moment of madness, having lived next door to the organiser while growing up, I'd joked about completing the 'Fred' in Antarctica, accomplishing the 112 miles on a stationary exercise bike. 'It would be fabulous publicity and I'm sure it would raise money,' he'd told me. In the Neumayer gym there were two bikes: a standard black gym bike with a digital number display, and a brand new, more sophisticated machine with a twenty-seven-inch high-definition television screen mounted in front of the handlebars. With individual rider profiles, ride history and built-in power meter and heart-rate monitor, it had been delivered on *Polarstern* with the main aim of monitoring the overwinterers' fitness levels throughout the period of isolation. Companies interested in space travel invest massively in teams living in Antarctic research stations that are manned over the winter. It's the nearest humans can get to space travel without actually leaving the safety of Earth, though some would argue space is safer.

Every month, our fellow nine overwinterers performed mental tasks and tested their fitness and Dr Tim took blood samples to be frozen and analysed back in Germany. I had expressed interest in taking part in the programme, especially with putting my body under enormous stresses caused by the extreme cold outside daily, but having thought about being wired up while trying to just survive outside, I'd decided not to. Instead, to keep fit, I just rode the bike.

I had huge ambitions for getting super fit while being at Neumayer and reaching a higher level of health than ever before. The extended periods of bad weather seemed the perfect time and since arriving I'd ridden a few times a week. The bike had routes to follow and competitors to race against and the idea of it was inviting, but it didn't take long for the novelty to wear off. 'Pyramid Mountain' was a twenty-kilometre loop and my 'go to' ride. Depending on what mood I was in I'd complete up to three laps, though usually I'd had enough after just one. I'd set a target time to get round and I did improve the more I rode, chasing down faster riders I'd built up a digital rivalry against. On occasions, Will, who loved running, would use the treadmill next to me. He also had ambitions to maintain his fitness and was a little more disciplined than I was.

As we headed into the winter, even after a few days off I was struggling to find the energy to ride; the outside was sapping all the energy that I had. The idea of completing the Fred on the same day as it happened back home in the Lakes was just a step too far. I was struggling with sixty let alone 180 kilometres. Being unable to freewheel down the hills meant I was putting in double the effort anyway. It was an idea I had to let go of, and it made me sad. I was missing home and the freedom of being on my bike.

The increasingly persistent winds had also started affecting our skidoos, which were living outside the front door. Although sheltered from the main force of the winds, they were still affected by the swirling mass of snow that the central column of the station produced. Being parked outside, I worried about the hammering our skidoos were taking. I'd already driven into a sledge, snapping its tow hook and putting a hole in the front of the skidoo, and having also snapped the suspension on another machine I was desperate not to give the mechanic more work to do. I questioned whether it was good for the machines to be sitting out in such low temperatures day in, day out, but it was our only option.

The skidoos' tarpaulin 'garages' proved a nightmare and poor protection in severe weather. It was all we had. The twelve tiny fabric straps and metal buckles around the base that secured down the main cover were not only a nightmare to do up, but loosened off under the lightest pressure. Being such small buckles, when I was wearing my large mitts they were just too fiddly to tie down and exposed skin would quickly freeze and stick to the metal loops. It wasn't just a pain each time I put my skidoo away, it started to become dangerous in temperatures reaching −40. I didn't want frostbitten fingers, so with cable ties and carabiners out of a spare mountain rescue box I replaced them all to make life easier. A simple large clip would surely be easier than threading a thin strip of fabric through small metal loops.

Temperatures below −40 started to become a regular occurrence and as we had only lost the sun a few weeks before, I began to wonder how much colder it was likely to get. Back home I had always associated the coldest months with January and February, so with Antarctica's version, July and August, still to come, it wasn't likely to warm up any time soon.

On computer screens in the offices, living room, kitchen and dining room, a large display showed real-time weather information as well as an upcoming forecast. Showing outside temperature (including a reading that took into account wind chill), wind speed and wind direction, it was an extremely detailed resource. However, every time I saw a temperature reading of −35 or lower, my heart sank. Working in such conditions was proving incredibly uncomfortable. The day Walter was born had been the coldest my body had ever experienced, but I was amazed at how quickly I'd become used to the severe conditions. When temperatures reached −40, however, I was suffering. It was excruciating. Will had experienced a bit of minor frostbite on his fingers, which had come as a big wake-up call to me, so I ensured I took extra care not to suffer the same.

Clear and extremely cold days meant I could get back down to the birds to continue my quest to film one laying an egg. It had been a few days since missing that first female lay her egg, and it was becoming a personal mission to capture one in the act. Temperatures hit the lowest I'd felt at −44, which felt like well below −50 taking into account wind chill. As they did on extremely cold days, the penguins had grouped tightly together, taking up a smaller area on the ice than they did when the temperatures were slightly warmer.

Immediately, it was easier to single out individual couples that stood around the periphery. It was still very much a case of learning as I went, as that first couple whose egg I'd missed being laid had given me very few clues on what to look out for. Keen to witness his first emperor egg being laid and to give me a helping hand, Will had come down to the colony with Stefan and me, his fingers wrapped up to keep them warm. Loading the camera and tripod onto my sledge, I started to pull it across the ice around the edge of the colony. The temperatures had solidified everything; every ridge and indentation left in the snow had become a rock-hard obstacle. My camera sledge skis scraped along the ice as its edges slid over mini mountains. Peering through the tiny horizontal gap between my hat and balaclava, I examined each couple that stood out from the main congregation. Ice had started to build up on my eyelashes and I could only just feel the bridge of my nose, the only area of my skin exposed. Even though I had a pair of clear goggles around my forehead, I didn't pull them down over my eyes. I needed my eyes free to be able to use the camera's viewfinder at a moment's notice.

As had happened when the eggs had been developing inside the females, pairs stood motionless, waiting, conserving every ounce of energy they could. Couple after couple, I surveyed the group closely for a few minutes looking for tiny nuances in behaviour that would give me a hint as to which birds were ready. Nothing. The more I looked around the colony, the more single males I could see standing with bulges already on their feet. A lot of birds seemed to be laying either during the night or out of my sight and I started to worry that capturing the main event was going to be more difficult than I thought. Despite it still being the early stages of egg-laying, I felt it was getting desperate.

Glancing into the main mass of penguins, all of a sudden I noticed a bird in a hurry, trying to get out on the far side. It appeared it needed some personal space. In close attendance behind was another emperor, following every step. Once out onto fresh ice they settled down together, standing so close their feathers entwined. Towing my camera sledge behind me, I made my way round the colony, ensuring I didn't take my eyes off them. As I walked round I thought about my kestrel in her nice cosy barn surrounded by bales of hay. Despite it being the opposite end of the Earth, she would have been doing exactly the same at this moment while preparing to lay. Settling down, I sat on the ice with my legs spread wide, the camera in between.

The two birds had well and truly isolated themselves from all the others, stopping about fifty metres away. I looked at them both as they stood side by side and for the first time from their appearance alone I could distinguish which was the male and which was the female. Following weeks of gestation, it was clear how much producing an egg had taken out of the female.

She also hadn't eaten for well over a month. She was thin compared to the male who stood tall and plump by her side, and she looked tired. She'd lost approximately one third of her bodyweight, which for any creature is an enormous amount. Other birds attracted by the couple walked over. They were usually graceful and social creatures, and this was the first time I'd seen the emperors feel uncomfortable around other birds. Protecting his partner, the male quickly saw them off, swinging his head around and swiping his long sharp bill within inches of the other birds. He circled around her as if creating an invisible bubble in which to shield her.

Showing her first signs of discomfort, the female began to edge further from the colony. Appearing to limp, within a few metres she came to a stop. The male seemed to know what was about to happen, becoming more and more attentive towards his hobbling partner. Again, the female attempted to move, but after three steps, gave up. She raised herself up high above the ice, exposing her feet and ankles; I could see her tail feathers pressing down against the ice through the clear gap between her feet. Usually, in trying to stay warm, the emperors stood hunched up, feathers almost covering their whole feet. It certainly looked unusual. Was she making room for an egg? I altered my focus onto her head. One single shot of an egg arriving wouldn't have been enough, so to help make an egg-laying sequence I tried to cover as much of her as possible. It was a risk: by moving the camera, I was in danger of missing the key moment.

Her eyelids closed slightly as she squinted in discomfort. Neck muscles elongated and contracted, exposing the outline of her breastbones under her thin chest. Scrunching her neck, she gasped heavily. The signs she was showing were identical to that of my kestrel and I knew an egg was imminent. To my surprise the male, swaggering off, swayed his body from side to side, heading for the colony. Quickly turning around and returning, it was as if he was bragging to the others about his female and what she was about to do.

She was continuing to show signs of pain; it was clear her contractions had started. Her tail bobbed, flicking snow between her legs and then back out behind as it returned to its normal position. I had my eye glued to my camera as the male re-entered the picture, his beak heading straight down towards the female's feet. He preened his white belly, and although periodically displaying normal behaviour he was just as restless as his female. His heavy winter suit weighed him down and walking around at times looked just as challenging for him as it was for the uncomfortable female.

All of a sudden the tip of the egg appeared between the female's feet, hanging with its pointed end down. This was the moment I'd been waiting for. I didn't dare blink. Keeping the camera recording, I watched in anticipation. With the egg visible, she adjusted her stance. As if to reward me for my efforts she turned and faced me, her feet displaying perfect symmetry in my viewfinder. Her toes curled inwards and as her tail flicked forward to cushion the fall, the whole egg appeared on the ice. I'd got it. Between her feet lay a perfect white egg. About twelve centimetres long and eight centimetres wide, it was huge, much bigger than I'd expected. Although I'd seen an egg being laid a few days before, this time, I wasn't frustrated and distracted that I'd missed the egg-laying moment, so was able to fully appreciate what I was looking at.

Immediately, both birds trumpet-called in unison, breath pouring out of their mouths as they seemed to celebrate together. Within seconds, to prevent it freezing on the hard ice between her ankles, using the tip of her long sharp beak, the female levered the egg onto the top of her feet, carefully covering it with a thick blanket of her white front feathers. At sub −40 degrees, it wouldn't have taken long for the embryo inside to freeze.

Even though the egg had arrived, the pair of emperors didn't seem to be settling down and it suddenly dawned on me that the next phase may happen faster than I'd expected. The male immediately started showing interest in the egg and without giving his partner any time to recover from her ordeal he started trying to take charge of it. Advancing with his head low, he nearly knocked the female off her feet. He was clearly keen to get hold of the egg but with the female appearing reluctant they eventually settled down, resuming their couple's stance next to one another.

I sat up and looked over to them. Having witnessed the whole episode on the small screen within my camera's viewfinder, I wondered what was going through their heads. Depending on how quickly the female would be willing to transfer her egg over to the male, these were potentially the last few moments they were to spend together until she returned in over sixty days. Did the male really want to take responsibility for such a precious item?

Since arriving back at the colony after the summer he'd been slowly shedding a small number of feathers on his belly just above his feet. This area of bare skin is present on all birds that incubate eggs, but is probably most obvious in emperors when they stand tall. Every few moments the male leant forward and, using his long curved beak, made final adjustments, nipping away any loose feathers that seemed out of place and ensuring his brood pouch was just the right size. The patch of pink bare skin on a bird so densely covered in thick black and white feathers looked raw without an egg hiding it.

Calling together, both birds stood tall, as if showing off their egg and brood pouches to each other. I wondered if they were signalling to one another that they were ready. The male again leant forwards, delicately scraping the tip of his beak across the surface of the precious egg. The female still seemed reluctant to relinquish her power over the egg. I zoomed my camera lens towards the egg and without warning she abruptly but carefully parted her feet, stepping backwards, exposing the fragile egg to the elements and freezing ice on which it lay. It filled my frame. The male reacted immediately, keeping the tip of his beak within an inch of the egg even when not actually touching it. He stepped forward, his eyes entering the top of my picture. With his wide-open eyes it was obvious he was aware of the importance of the moment but he didn't seem to panic; he had to be quick enough to prevent the egg from freezing, yet not so quick that he risked cracking it. Using the very end of his beak, he gently dragged the egg between his feet, the point facing away from him, and in a final manoeuvre cautiously brought his feet together, lifting the egg upwards and away from the lethal ice. Together, they'd successfully completed the most risky part of their breeding process and I could breathe a sigh of relief. I had got it.

The only thing left to happen now was the female leaving to head out to sea; the process had been incredibly quick and I readied myself for the final act. It seemed like role reversal; from the male paying extreme interest in the female's egg, it was now the female that paid interest in the male's egg. It wasn't that she wanted it back, she just wanted confirmation it was safe and that she could trust him before she disappeared. He slouched, lowering onto his feet, blanketing his egg with his feathers. Dropping his shoulders, he closed his eyes. As he shivered in the cold, the male knew this was it and the female got the message. He was ready. She turned around. Walking away, she momentarily hesitated and glanced back. I couldn't help but translate it as, 'Stay safe, I won't be long.' Her head dropped, she fell onto her belly and started to toboggan away into the distance.

The egg was now the sole responsibility of the male through what were arguably the toughest conditions on the planet. As the wind picked up, the female vanished into the snowdrift as she made her way north in the direction of open water. Weak and hungry, she had no idea how far her journey would be and I was amazed she still had the energy.

I sat back and turned to Will, who'd been at my side watching the whole process as I'd filmed. The whole courtship ritual and mating process, which had taken almost two months, had concluded in an incredibly short but significant moment. The process had taken under forty-five minutes from start to finish, but having had my own eyes buried in the camera capturing as much detail as possible, what I'd just witnessed hadn't quite sunk in.

The strength of the bond between the two birds was overwhelming and, all of a sudden, I was struck by the parallels between a pair of emperor penguins and my relationship with

Becky. The pair had just separated, not knowing whether they'd ever see each other again, the female unaware of whether their egg or its father would make it through the winter, and the male unsure if his partner would survive two months out at sea. As I thought of the moment I'd left Becky and our unborn baby at the front door of the house, disappearing round the corner in the car, it hit me. Witnessing moments such as a pair of emperors laying an egg, transferring it between them and saying goodbye to one another was why I'd travelled to Antarctica, why I'd sacrificed such a huge amount. I was seeing natural events that only a handful of people could say they've ever witnessed, and these were the moments I'd dreamt about. Tears filled my eyes and froze immediately to the top of my cheek and balaclava; clearly, my emotions were very close to the surface.

I looked at the male and knew how he felt. I knew the bond between the two birds was strong but I had no idea it was *that* strong and I just wanted to go over and give him a hug. For the next sixty or more days he would be a single parent, battling to survive the darkest, coldest and windiest winter on Earth while at the same time caring for a fragile egg that balanced on top of his feet. The prospect was simply one of the wonders of the natural world.

Over the following couple of weeks there was a mass exodus of females as they laid their eggs, passed them over to their partners and left for the open ocean to feed. I needn't have worried about not managing to film an emperor laying as it didn't take me long to spot the obvious signs that expectant mothers showed. But they were never as obvious or as moving as that first couple that had so clearly showed me how it was done. Looking out towards the northern horizon, the sky remained clear and the air cold. Mirages rippled above the ice as the black specks of single females melted into the distance.

The colony was getting noticeably smaller. The cold forced the now egg-laden, less mobile males to shuffle closer together for warmth, and with all the females leaving, the colony had halved. Having started with almost 10,000, it was whittling its way down to 5,000 extremely quickly. New males that had recently taken charge of their partner's egg slouched down into their energy-saving winter posture. It was a stance that I'd associated with courage, bravery and bloody-mindedness; what the males were about to endure was beyond comprehension.

As the males inched forward towards the group, taking their first few steps with an egg balanced on their feet, I wondered how on earth they'd manage it. It was a fabulous time to be with the colony; days were extremely short but the behaviour they were displaying made it so exciting to be there every day. The contrast to just a month before was huge. The colony had been a rowdy mass of males and females all calling to one another with the aim of attracting a mate, but with pairing up and egg-laying completed, the males had no reason to be wasting valuable energy in vocalising. The group fell silent, just the odd squawk from a restless male occasionally being heard.

With yet another bank of cloud heading over from the west, filming the egg-laying came to a close. Very few females were left in the colony and I knew by the time the weather had passed and I was able to return to the birds there would probably be none left at all. I felt proud about what we'd managed to capture, and even though our time with the birds had been tight, the weather we were filming in had been kind to us and I hadn't felt in danger at all.

One of the classic images of Antarctica I'd grown up seeing, especially of the South Pole, which lay 1,500 kilometres to the south deep in the heart of the continent, was one of a signpost showing locations, their distance from Antarctica and in which direction they lay. I'd been keen to make one to take home with me and with the long dark nights, I set to in the workshop. As well as one with my home town on, I made one for Walter. In an old-English-style font I stencilled out my location, his name and his birthdate underneath an outline of an emperor and its chick. Using Neumayer's saw bench I cut out the shape of the sign on a sheet of three-quarter-inch Baltic birch plywood. The station had a huge stack of wood for various maintenance projects and it was a strong, heavy material that looked like it could put up with the outside conditions. Over the course of a couple of evenings I engraved Walter's name in the wood, sanded it down and gave it a layer of varnish before taking it up to the generator room to dry. Each evening I'd inspect it and return to the workshop to paint the letters black and shade in the penguin and her chick. With obviously no shops at Neumayer, I couldn't purchase any gifts for Walter, so my intention was to take it down to the colony, take a photo with some emperors behind it and bring it home with me.

Early one afternoon I went into the office to view some egg-laying footage that Will had edited together and to sort out some kit. As the temperatures had been dropping, for the first time I'd been experiencing problems with kit and various other items. Having been busy, I'd not found time to repair them. The black plastic casing of a long cable that ran from my tripod handle to the camera lens had snapped into several pieces, exposing its metallic core. As it was a complex cable it wasn't something we could easily replace so I'd covered it with green electrical tape. I had a couple of spares in our kit room but was reluctant to break into them until my original cable truly gave up.

The small rubber button that my thumb pressed against to start the camera was freezing rigid in temperatures below −40 and made recording at key moments challenging. Luckily, with Neumayer's array of various technical scientific equipment, parts such as plastic buttons were in abundance. I gave Stefan the job of drilling into our extremely expensive piece of kit to exchange the button.

The camera batteries were holding up surprisingly well in the cold and the only difficulty I was experiencing was changing them once they died. Shutting down the camera, releasing the battery that clipped into its rear, replacing it with a new one and turning the camera back on was proving painful as I'd have to remove my mittens and the process was extremely time-consuming. By cabling together four batteries we produced a unit that plugged into the back of the camera, powering it for a whole day without having to worry about it running low. The four batteries sat snugly in a small insulated bag we'd had designed and custom-made back in the UK.

We also discovered that we needn't have spent a fortune on the special thick blue Arctic-grade cable. Much thinner flexible silicon-coated cable didn't respond to the cold in the same way that all our other cables did and thankfully the station had an abundant supply we could borrow. Everything I did was proving to be one giant learning process and many of the plans we'd made prior to travelling were proving useless. The conditions dictated what we could and couldn't do and each day taught me something new.

I rarely brought the cameras inside the station as the difference in temperature between outside and inside was so great that I risked breaking them. Thermal shock on the glass of the lenses was a real risk and a build-up of condensation would soak the electronics inside. It was just too severe to be doing it at regular intervals so for the majority of the time they lived in their sealed boxes on the back of the skidoos. One evening, having fixed everything that needed our attention and packed it up ready for when the weather broke again, I opened the boxes that the cameras were being kept in. This was the first time I'd attempted to bring my new set-up inside since swapping to the low-light cameras, and as the weather was forecast to be bad for at least the next four days I'd taken my time, warming the cameras extremely slowly. It had taken me three days to get them from −45 outside to our +20 degrees Celsius office room where I wanted to clean them, completely dry them out and make sure they were still working as they were meant to. Each day I moved the cameras in their boxes, surrounded by socks filled with silica gel, up a floor, gradually exposing them to the warmer heated air of the station.

I had two cameras, one with a wide-angle lens and one with a long lens, which I used for the majority of the work with the penguins. Having two was a lifesaver as changing delicate lenses in swirling snow just wasn't possible. I opened the two boxes but both cameras appeared wet, droplets of water covering the lens, body, battery and viewfinder. Without thinking, I attached a battery to the back of one and flicked the 'on' switch.

Immediately the viewfinder flashed several times. For some reason I stupidly repeated the process with the other camera. The same thing happened. I'd not had a camera behave like this anywhere before, let alone in a place where it's not possible for them to be replaced. I kept calm and quiet, not saying anything to Will, who sat next to me working away. I stripped both machines, removing everything I physically could, and opened every door and socket they possessed. Taking them down to the warm generator room I laid them down on a dry towel to completely dry out, my sign for Walter still propped up against a warm pipe.

Despite being deafeningly loud, the enormous generator that was powering the whole station produced a huge amount of heat and it was the perfect place to get things dry quickly. Quietly, I was panicking, worried that I'd accidentally damaged both our winter cameras beyond repair, the only two cameras we had that worked under such low-light conditions. With still well over a month before the sun was due to return, when I'd be able to go back to my original set-up, I thought about how I'd tell the team. I'd been so careful and, other than a few cables, not broken anything. Turning off the generator room's light and closing the heavy door, I prayed they'd come back to life. I checked weather forecasts to see how long I could leave them for; with a build-up of moisture inside, the longer I left them to thoroughly dry out, the better. White-out conditions were expected for at least three more days so I made sure I left them well alone.

Trying to keep my mind off the cameras, I was keen to see how far the female penguins were having to travel once they'd laid their eggs. Using an online service, I checked ice conditions. Radar images of Atka Bay taken from space were updated every few days and I could see how far the sea ice extended, where icebergs lay and any cracks that had formed in the ice. One thing I was desperate to see up close was the emperors swimming and leaping out of the water onto the edge of the sea ice. With unpredictable weather, winter wasn't the time to be trying to get to the edge, but seeing how far we'd have to travel gave me ideas for when the weather did improve.

I downloaded the most recent image and zoomed in on the area our colony had congregated on. Straight north, with nothing blocking their route, I drew a line to where the nearest patch of water lay. The females didn't need vast wide-open expanses of water, just somewhere the ice was less stable and broken up that presented holes through which they could come and go. Their route measured roughly thirty kilometres, a fair undertaking for any animal to attempt, let alone one that hadn't eaten for over a month in the coldest conditions on Earth.

Now I'd been away from home for well over 150 days, life had become simple and straightforward. I'd had to become self-sufficient in certain ways. I'd cut my hair three times with some clippers I'd brought with me. I was no expert but a 'number seven' all over was working fine, tapering the back and sides with numbers six, five and four. It didn't look too bad. Will had requested the help of two of the scientists to cut his hair one evening and it hadn't gone well. I was pleased I'd made the choice to do mine myself.

As with all male intrepid explorers, I'd given growing a beard a go. I'd never managed more than a couple of millimetres back home before Becky expressed her distaste and I'd shaved it all off. Being 15,000 kilometres away from home, however, I'd let it go and to my surprise it was curly and bright ginger. It felt extremely uncomfortable and was not attractive in the slightest. Having persevered while it froze to the inside of my balaclava when outside, I had finally had enough and took the clippers to it.

The same evening, sitting in the office, I scratched my neck where part of my beard had been growing. As my fingers glided down under my chin they ran over a lump on the left side of my neck. I was no doctor, but a lump immediately alarmed me. It was only the size of a small marble but I'd never felt anything like it before. I'd always been one for worrying but being isolated and out of reach of help, everything felt ten times worse. So many things went through my head. What could it be? Would I need evacuating? I sat on it for a couple of days worrying without telling anyone and then arranged for Dr Tim to take a look. At Neumayer, the station leader doubles up as the doctor but isn't required to do anything else. Therefore, when someone requires medical attention, they can focus on the issue without any distractions. Quickly tilting my head back, he checked it out.

'It's absolutely nothing to worry about,' he exclaimed, 'just an ingrown hair.' Immediately I felt relieved. I told him how silly I felt having been panicking for a few days. Everything, no matter how small, felt massive in Antarctica.

Once a month, Dr Tim arranged a day for fire drills. It was important we continued to practise and refresh the skills that we'd all learnt on the firefighting courses back in Germany. A fire on the station would have been disastrous so the doctor made sure everybody was present when drills took place. Even though we knew the day the drill was due to happen, he still surprised us when the alarm sounded. The gathering point was in the hallway outside the cloakroom and working in pairs everybody had their own jobs; one pair would act as search and rescue, with another pair backing them up. Our IT technician was in charge of radios and keeping a record of times and oxygen levels for those wearing tanks. We all wore our own red overalls with reflective strips around the ankles and arms and had our own personal oxygen masks. We took each drill very seriously and sat around the dinner table with coffee to debrief after each one, discussing where we had gone wrong and what we could have done better. We were prepared and had kit available to tackle any size of blaze and we'd even filled emergency bags with fresh warm clothing, spare shoes and over six weeks' worth of food for the twelve of us.

When the last plane had left, these bags were relocated to the E-Base (or emergency base) two kilometres north of the station, positioned along the same track that led to the Northern Pier. The E-Base was a set of orange shipping containers raised on stilts, just like Neumayer III but on a miniature scale. I certainly never expected to have to make use of the E-base, but it was comforting knowing all these measures were in place if something serious were to happen.

One afternoon, sitting in the office discussing filming with Will, a loud yell of '*Scheisse*!' followed by the fire alarm sounding came from just down the corridor. Rushing to my feet, I immediately knew it wasn't a drill; the doctor always told us when they were happening. Stepping out from the office into the hallway, I looked down towards the meteorologist's lab. He'd been sorting through a mountain of large lithium ion drill batteries, which he intended to use to monitor sea ice thickness. One of the batteries, positioned away from the others, lay in the middle of the floor making a quiet hissing sound while yellow smoke poured out of it. Panicking, the meteorologist had smashed the small red alarm box next to his office door.

Will and I ran past the battery down the stairs to our meeting point where everybody was kitting up. It was a serious situation and we had to be quick to take control of it before it got out of hand. Just as we went through a register to ensure everybody was present before sending people back to deal with it, the geophysicist announced the danger was over. Grabbing the nearest fire blanket, he had wrapped up the smoking battery, taken it to the nearest fire escape and launched it out of the door onto the ice. It wasn't what we'd trained for or how to deal with a fire, but the crisis was over and we could relax.

We weren't sure what had happened with the battery but while smoking it had produced so much heat that it had melted a small area of the floor. It was a huge wake-up call and we'd been exceptionally lucky. I looked over at our desk in the corner of the office. Our camera battery charging station was stocked with a mountain of batteries and casually we'd been recharging them all day, every day. Our kit was taking a hammering and even though we were careful to bring them back up to room temperature each evening, it dawned on me how dangerous they could be.

With a more favourable forecast for the following day I returned to the generator room to rebuild the cameras that had been drying out since they'd become covered in condensation. I had no idea how they'd respond when I attached a battery and powered them on, but as I did so I closed my eyes and hoped. Both fired into life quicker than they had ever done before. I can't describe the feeling of relief that came over me. Throughout the few days they were in there I hadn't stopped thinking that if the cameras were damaged beyond repair then our winter's filming was over until the sun returned. As winter was such a key period in the emperors' lives, we wouldn't have had a film without it. It was almost like the cameras were crying out for a break from the cold and were all the better for some much-needed TLC. I was hoping this was an omen for the remainder of our filming.

5.

The Worst Winter on Earth

As I entered the month of June, I hadn't see the sun for over two weeks. Despite the fact that the sun had disappeared below the horizon, I was surprised at how much twilight there was over midday, especially under cloudless skies. Not only was it dark for much of each day, it was unimaginably cold and for most of the time, exceptionally windy. The brutality of an Antarctic winter had well and truly kicked in. Just venturing outside caused my eyelashes to freeze together, making it difficult for me to even see, but for the penguins it was so much worse; they didn't have the comfort of a warm research station to return to each evening.

This new season altered the world outside; everything had been transformed into a mysterious murk and nothing looked the same. The extended darkness with no light pollution meant I could see millions of stars; the night sky extended high above my head and I felt I could see further than I ever had before. It just went on and on. Ice crystals on the frozen ocean sparkled under the faint light from the moon. In such a pure white landscape the moonlight transformed the scenery. There was so much beauty amongst the weather's brutality. The sun wasn't that far under the horizon but it didn't offer us any warmth or respite. The only reminder that the sun still shone elsewhere on Earth was the reflection from the moon that gave a little bit of light to the landscape in front of me.

Looking down on the snow at my short, dark shadow, I thought about the long summer days that seemed so far away. My family kept sending me pictures of the beautiful summer they were experiencing at home, the fresh green grass, the butterflies fluttering; it seemed a million miles away. The temperature difference between there and where I was standing was up to 80 degrees Celsius at times. I was in the middle of an Antarctic winter at the opposite end of the planet to everyone I loved and the feeling of being comfortable and warm outside was something I simply couldn't remember.

Physically and mentally, things were starting to get tough. It was crucial that I captured the full extent of an Antarctic winter to show just what the birds go through to reproduce each year. I started to feel the pressure. I owed it to them to show everything they went through; the cold and the winds are almost impossible to describe but maybe the pictures could do the experience justice and show just what they have to endure.

Male emperor penguins brood their large single egg for approximately sixty-four days, the longest continuous duration in the bird world. It's an epic feat. All day and all night they were having to battle just to keep themselves warm and to ensure their precious egg remained in position on the top of their feet. Each penguin was hanging on; the lives of the next generation depended on them. I started spending more time with the colony during hours of darkness. Although these were the worst conditions I had to go through, it was actually an extremely special time. It felt such a privilege to be with the birds during the most difficult time of their year. I felt as if we were in it together; we all had a challenge ahead of us. Roughing such brutal temperatures and storms was what emperor penguins were well known for; for me, on the other hand, everything was new.

With no sun to give any warmth, temperatures consistently hovered around –40 degrees Celsius on clear days. Every part of my body ached; my fingers and toes went numb, the skin between my eyes stung as if it had been burnt and my joints throbbed from hours spent kneeling on the solid ice. For the penguins, whose coat consisted of over a hundred feathers per square inch, making it the densest coat of any bird, tolerating these freezing conditions was what they had evolved to withstand. But that didn't make their lives any easier. Even for a creature so well adapted to cope with such extremely cold conditions, the freezing air eventually became too much. The only way they were able to get through it was to help each other. Before my very eyes they started to form a huddle. Huddling together was one of the pieces of behaviour that the emperors are most famous for and the behaviour I had been most excited to see. I'd been surprised at how low the temperature had dropped before they were no longer able to withstand it alone.

It all started with just two birds leaning on each other. Like a slow rugby scrum, emperors shuffled up behind and, linking in, they squeezed their fat bodies into the tiny gaps between two birds. Within just a few minutes, hundreds of penguins stood together, each with their beak nestled in amongst the neck feathers of the bird they leant on. Packing so tight, up to ten birds occupied each square metre of ice. For such a large creature, I couldn't believe how little space they took up when in their huddle formation.

Along the outside edge I watched a bird creep slowly along, stopping every now and then to inspect the gaps. He was looking for one just the right size to wedge himself into. Rotating his feet ninety degrees, he leant in against the wall of penguins and tucked in his beak to make himself comfortable. As they slouched over their eggs, all the male emperors appeared fluffy and fat, but as I watched this penguin settle down, his appearance changed. Like a miniature Mexican wave, his feathers flattened from the top of his head to the tip of his tail. Slowly, row by row, each line of feathers folded down on top of the one below. It was a minute detail and I had to look hard to see it.

As different birds joined the huddle, all smoothing their feathers, I considered why they were doing it. As a whole colony, the penguins were beginning to function together, relying on one another to keep warm. Rather than single birds keeping their own eggs warm, I was witnessing the creation of a giant incubator, brooding approximately 5,000 eggs, which unbelievably kept them up to 70 degrees Celsius warmer than the outside temperature. As well as physically doing all they could to conserve what energy they had for themselves, each male emperor was having to transfer some of that energy to his precious egg, in the form of heat. Their urge to breed and what they had to go through to succeed was incredible. Their aim was to move into the middle of the huddle and be surrounded by as many warm bodies as possible. To maintain a body temperature similar to that of a human, around 37 degrees, the male emperors were displaying one of the most incredible examples of working together in the natural world.

I looked across the seemingly motionless mosaic of row upon row of emperors. It was silent and with not a breath of wind the full moon began to rise in the east. Unlike the harsh bright summer days that had proved too blinding for my eyes without the use of tinted glasses, the atmosphere was soft and subtle. Even though it was lit by a bright moon, my pupils had been forced wide open to allow in as much light as possible. It was mid-afternoon and the sun had once again fallen so far behind the Earth that not a hint of colour was evident on the northern horizon. The single crowd of around 5,000 males occupied a tiny portion of sea ice. In contrast to when the colony had consisted of both males and females and when temperatures were warm enough for them to be independent, the surface area of ice they occupied was a fraction of the size.

With the emperors locked tightly together in a seemingly stationary position I began to notice small shifts. As one bird readjusted the position of his feet he released a small amount of pressure from his neighbour. This provoked a reaction causing the adjacent bird to do the same. In turn, a ripple effect flowed through the entire colony. The silent atmosphere was softly ruptured as the sound of scaly feet and claws shuffled across the ice. The whole group shifted just a few inches. I wondered just how effective their strategy was. The temperatures were so low that I couldn't imagine anything that would work to overcome them. I felt like creeping up and joining the huddle myself just to sense what the penguins were feeling. If only!

For hours I split my time between kneeling and sitting directly on top of the ice. The birds seemed settled and I tried to concentrate on capturing little moments of normality that individuals displayed. Whether it was an emperor coughing or simply blinking, they were all valuable images that would bring the unique motionless behaviour of huddling to life. My kneecaps were pressing hard against the ice and were struggling with the pressure my body was putting on them. After ten or so minutes I swung them round and placed both my outstretched legs either side of my tripod. Within minutes my backside had become so numb that I rearranged myself back onto my knees. It was a process I'd mastered! Peeling away the fabric of my polar suit, which had frozen solid and stuck to the ice, I was scared I'd rip it and make a huge hole, allowing in freezing air. I tried squeezing my hand into a tight fist in my thick mitten and repeatedly jogging on the spot, but there was nothing I could do to warm up my body.

Looking at the penguins that had managed to get themselves into the centre of the huddle, I wondered how they were feeling. Almost instantly a male lifted his head and started showing signs of discomfort. Surrounding birds immediately reacted in a state of panic and within seconds every penguin in front on me had begun a slow stampede away from the huddle. Each shuffled with urgency into their own space to avoid being knocked over. The risk of being crushed looked very real. I slung the camera over my shoulder and started to retreat. Birds were heading straight for me and I wanted to give them space to run into. Unbelievably, even in temperatures as low as −50, the single male in the centre had overheated and forced the entire huddle to explode so he could cool down. As if the ice was on fire, a mass of steam from the warmth the penguins had generated rose up into the moonlit sky. If I wasn't already convinced of the efficiency of huddling as a way of conserving heat, here was my answer! The penguins had, for the time being, turned off their giant oven.

I felt confident with the footage I'd filmed of the emperors huddling in calm weather but this was only half their story during the winter. My next challenge was to capture the emperors fighting against the full force of an Antarctic storm. Battling a storm had the potential to be a defining moment in the film and I felt that, without this sequence, I'd not be telling the full story. Winter conditions in Antarctica are well known to be the most atrocious on Earth and surviving such conditions was behaviour I felt I simply had to record. Every day I spent at the colony the iconic *Planet Earth* footage I'd seen as a teenager, which had done so much to inspire me, played out in my head. Out of all the behaviour I expected to see, experiencing these conditions alongside the emperors excited me the most. It seemed crazy but it was that ferocious footage that had attracted me all those years ago, and this was my chance.

Having entered the quiet period during which the males were only incubating and huddling to keep warm, I started trying to plan a trip to film the birds in bad weather. Although I'd experienced countless days when I'd been prevented from reaching the colony due to unsettled weather, the penguins had been relatively lucky so far with the conditions they'd come up against. The storms during the autumn had never reached a violent, life-threatening status and I hadn't yet seen any casualties. Safety in such a dangerous environment was a huge issue for me and a lot of aspects had to fall into place to ensure we returned to the station healthy with the footage we were after. I knew I wouldn't get many chances so devising a plan for when an opportunity arose became paramount.

Trying to reach the colony of emperors *during* a storm that had already begun would have been quite frankly stupid. Not only would it have been impossible to find the birds on the sea ice in zero visibility, it would have been extremely dangerous not knowing exactly where we were on the frozen ocean. I started looking closely at the forecast our meteorologist at the station was producing each day. What I was after was a weather system that would hit around the hours of midday, enabling me to be in position next to the birds as the winds picked up. As I was still in the time period when the sun wasn't rising, I needed to film the storm during the brightest hours of twilight, around lunchtime, to give my camera enough light. For almost ten days, the air was clear and calm, and despite giving me great opportunities to build on my huddling sequence I started to worry that I wouldn't get a chance to capture the males in a storm before the eggs began to hatch. I had mixed feelings; I was desperate for the birds to make it through incubation in comfort, but on the other hand I was keen to experience and witness what they had to do to survive such severe conditions.

All three of us were keen to see the birds reacting to a storm but Will, Stefan and I were equally keen to make sure we did it safely. Working in potentially life-threatening conditions, we couldn't afford to have any tension between us. We had to all be in agreement. Filming in a storm was a great worry for Miles and the team back in the UK with whom we'd been keeping in close contact, updating them of our plans. Despite them trusting us entirely they couldn't help but feel responsible and wanted to know we were doing everything we could to remain safe.

Over dinner one evening our meteorologist read out his forecast for the upcoming forty-eight hours. The nightly ritual of him predicting the seemingly unpredictable had become a running joke. He'd built himself a reputation for forecasting storms when actually calm weather arrived and predicting clear skies when clouds shaded the station. I felt for him. Being on a remote Antarctic ice shelf couldn't be the easiest place to accurately forecast the weather. When it went wrong and our filming schedule was affected he felt like it was his fault and I did my best to make sure he knew it wasn't.

As usual, that evening, he predicted the precise wind speeds he was expecting: up to forty-four knots were due to hit around the following lunchtime. He may have got it wrong on numerous previous occasions, but if he was right this time, the conditions seemed perfect for a first attempt at filming the birds in some serious weather and it felt worth a go. The line between success and failure was fine. It had to be windy enough to make the conditions look uncomfortable but not so windy that being on the sea ice would be dangerous.

The next morning the three of us spent half an hour kitting up into our polar clothing and jumped on our skidoos to head down to the penguins. As I closed the station door behind me I felt excited but slightly nervous. I knew that if the storm came in as predicted I'd be experiencing the most physically demanding day of my life. I looked up at the station and wondered when I'd see it again. The sky was overcast but remarkably bright for a winter's day and the wind speeds were still low. Travelling down along the flagged route was like any other day and, on arrival at the colony, the birds were clearly feeling the cold having packed tight and re-formed into their compact huddle. In a north-to-south direction the heads and shoulders of the 5,000 penguins stretched across the ice in a rectangular formation. As I set up the camera equipment a grin stretched across my face, hidden under my balaclava; if the forecast was correct, I was about to witness events that only a handful of people had ever had the chance to see. Everything was crossed.

Bang on midday, long wispy lines of drifting snow started appearing across the sea ice from the east. Like snakes, the slithering lines of ice crystals blew ever closer as the breeze started to increase. Our weatherman had got it right! The exposed penguins on the outer edge of the huddle quickly became restless, as if they knew what was coming. In no time at all the light breeze had turned into a blizzard. The slithering snakes had transformed into herds of buffalo as chunks of ice were blown through the air above head height. Visibility rapidly reduced, engulfing the majority of the huddle, and the occasional strong gust that came through reduced visibility further to practically nil. I looked behind me towards Stefan and Will, who were standing next to the skidoos. It immediately became clear that venturing any further from the skidoos would carry the risk of not being able to find them again. The wind was deafening even under my hat and balaclava.

The penguins responded to the onslaught immediately, with the majority burying their heads deeper in between the two males they leant on. But as a large bank of snow built up against the edge of the huddle the penguins knew they had to do something to prevent themselves being buried alive. Breaking away, they started to make their way around the edge to the more sheltered side. Exposing more birds to the rapidly strengthening winds forced more to peel off and seek the shelter of the protected side. In one long single file, males shuffled around the outside of the colony feeling the need to rejoin the huddle out of the wind. Like a conveyer belt, the emperors rotated around the group, trying to protect themselves. The colony was on the move; the longer the storm roared, the further it travelled. Every few minutes I grabbed my camera and moved to keep up with them, making sure I didn't travel out of sight of the skidoos. Will, Stefan and I had agreed beforehand that they would be in charge of safety, freeing me to concentrate entirely on capturing the events on camera.

With a constant queue of emperors around the edge of the colony, ridges of snow either side of their track began to resemble small mountains. For the male emperors, all of whom were still cradling their treasured eggs, these formed potentially deadly obstacles. One by one, each male negotiated the ripples across the ground. Shuffling their feet up and over an inch at a time, they carefully traversed the solid mounds with their eggs precariously balanced on their feet.

I tilted my camera towards the next bird to run the gauntlet. Impatient penguins behind started to stack up like dominoes, and the line of emperors fell forwards. The lead emperor was forced onto his belly; this was disastrous. He was holding onto his egg for dear life. I immediately looked towards his feet. I was convinced a fall like that would have forced him to spill his egg onto the ice. His tail, tucked in between his feet, appeared to be wrapped around the egg: the only thing that had prevented it from falling out. Using his beak as an ice axe, he dragged himself forward out the way of the impatient queue of birds behind. A strong gust of snow blew through, separating me from the penguins. For a few seconds they vanished behind it.

Looking away from the camera, I was forced to kneel down with my back to the wind and cover my eyes. The snow and ice particles being pelted against my face made the skin over the bridge of my nose sting. Ironically, it felt like it was having boiling water poured onto it. To relieve the top half of my face I pulled down my clear goggles that were around my hat and placed them over my eyes. They'd already filled with snow but the immediate relief I felt out of the wind reminded me how severe the conditions had become. My mind was focussed only on capturing the behaviour in front of me.

The gust passed and I stood up, but the emperor was still on his belly desperately trying to right himself. Without being able to use his feet to their full effect he dug his beak into the snow and pushed his neck upwards. Creeping his claws forward, he slowly arched his back until his balance returned him upright. It was incredible to see how flexible the male could be without letting his precious egg slip away. The storm continued to intensify and despite it becoming almost impossible to stand up against its strength, my excitement grew. Adrenaline was pumping through my body as the behaviour that was unfolding in front of my camera was truly unique. Even biologists who've studied emperors for years rarely get the opportunity to observe the birds in such callous conditions.

Lifting the kit to reposition once again, I slung it over my shoulder. The wind immediately caught the camera like a sail and swung me round. At over twenty kilograms the tripod and long lens sitting at head height had made my body top heavy and had put me off balance. Slamming the three tripod feet back onto the ice, I just managed to save myself and the kit before we tumbled onto the ground. In wind peaking at close to 100 kilometres per hour, falling over with the camera balanced on my shoulder could have broken my neck. Realising the danger, I immediately began to think about my safety. Will had seen me nearly topple over and, leaning into the wind, he approached me to see if I was OK. The gusts had picked up so much that the colony was almost invisible behind the wall of snow blowing through.

'I think we'd better leave,' shouted Stefan. All agreeing we'd outstayed our welcome, we decided to pack up and attempt to get back to the station. I was keen to continue but understood we had to make a move. The weather was only due to deteriorate and, having recorded some extraordinary images demonstrating what the birds were having to put up with, we had to prioritise our safety. As the colony rotated and began to disappear into the heavy drifts of snow, I let them go.

Having withstood the storm for almost four hours, we started to leave. With Stefan's skidoo still running, Will and I tried to turn ours on. I wondered how much snow had accumulated in the engine. We'd experienced problems in heavy blizzards before, with the skidoos struggling to restart, so having one running was an insurance for the other two. After a nervous twenty seconds, they fired into life, their large headlights illuminating the barricade of snow that blew across in front of us. Over time, because I would choose which part of the ice I wanted to film from each day, I'd frequently been the lead in our line of travelling skidoos. So now, I set off in front with Stefan close behind me and then Will at the back.

Immediately, I knew the journey wasn't going to be as straightforward as I had thought. Had we left it too late? I'd dialled Neumayer into the GPS mounted on my handlebars and the route along the flagged track appeared. It was eight kilometres away; under normal circumstances, the trip would have taken around ten minutes. For our individual safety all three of us had our own GPS units, but we stuck close together. As we drove up the ramp of snow off the sea ice onto the elevated shelf we stopped and made sure we were all happy to continue.

As we set off my eyes were glued to my GPS unit. A red arrow symbolised my location and a blue line represented the preplanned route I was following. All I had to do was keep the red arrow on the blue line. Visibility had reduced so much that, without Stefan's front headlight, I wouldn't have been able to see him. Unable to see any further, I was relying on Stefan to keep an eye on Will. With hardly any light left we were under pressure to get back to Neumayer before conditions worsened.

Slowly, I accelerated, creeping forward and ensuring I didn't deviate off the route. There were potentially deep crevasses in the ice on either side of us and as I couldn't see the snow on which I was travelling, I was completely dependent on the little colour display that sat in front of me. The wind was becoming so strong that I was being blown off my seat. To keep my body upright I put almost all my weight onto my right leg and pushed so hard it began to cramp up. With my goggles full of snow I had no option but to expose my eyes and forehead to the elements. I glanced into my wing mirror then back to my GPS. The only sign that Stefan was behind me was the faint glow from his headlight, which vanished momentarily behind gusts of blowing snow. With wind chill, the temperature had fallen well below −50 and we were still over six kilometres from the safety of the station.

Edging forward, I didn't dare accelerate; the track hadn't been ploughed for a few weeks and large mounds of sastrugi had formed and frozen solid across the route, making our journey bumpy. I looked in my mirror, back at my GPS, back into my mirror, then back to the GPS. All of a sudden, the faint glow from Stefan's light had vanished. I stopped immediately and turned around on my seat, shading my eyes with my thick mittens from the icy bullets that were blasting my face. Nothing. I could barely see the box strapped to the end of my sledge, let alone another skidoo further back. My instinct took over as I grabbed my radio from my top pocket. I crouched beside my skidoo to protect myself from the full force of the wind. Before attempting to make contact, I realised the three of us were on an open radio channel. Announcing we were split up over the radio would have alerted the guys back at the station, and I didn't want them to know we had got into difficulties just yet. I held my nerve and waited. Five minutes went by. My mind started to race. Where were they? Were they still travelling together using their own GPS units? Were they even together? Should I wait or turn back? The GPS units we were using were identical and we all had the same routes displayed in front of us. I was trying to remain calm but I began to worry. I'd been monitoring my mirror with such concentration I couldn't work out how they'd disappeared so quickly. I looked back down at my GPS, where a red zigzagging line on top of my blue track displayed the actual route I had taken.

Feeling I couldn't wait any longer, I decided to retrace my steps and go and look for them. I quickly spun my skidoo round and began to follow my route back. Within literally a few metres two headlights side by side appeared out of the blizzard. Crouched between them were Will and Stefan. Just like mine, Will's goggles had filled with snow and iced up. In such freezing and blustery conditions, the loose particles of ice blowing through the air got everywhere. Like sand, it filled pockets, squeezed through zips and penetrated the thin foam membranes around our goggles. Stefan had surrounded his with a layer of sticky tape before we'd left the station that morning and kneeling down next to Will he attempted to help. Having taken his gloves off, Will was trying to scratch the layer of ice off the inside of his goggles with his fingernail.

'Are you OK?' I shouted over to them. I worried about Will, who had already experienced minor frostbite on his fingers. 'We're not that far now, try and do without the goggles,' I suggested. Already almost an hour into our trip, we re-formed our single file and continued. To make sure I didn't lose them again I travelled even more slowly. The conditions were easily the most terrifying and dangerous I'd ever experienced, but despite feeling scared, I found it exciting. I always knew in the back of my mind that the safety measures we had in place and the training we'd completed for this very scenario would see us right.

Even though we stuck together, all three of us were individually prepared in case we became separated. We each had our own radio plus a spare battery in our top pockets for easy access. Mounted on the handlebars of all three skidoos were GPS units. We were also carrying in a separate pocket an additional GPS unit, again with a spare battery. As the ultimate safety precaution, we carried with us a GPS tracker, which enabled the team back at the station to view our exact coordinates at any time on a piece of computer software. Bolted down onto our sledges, in addition to the mountain rescue equipment, was a large red aluminium emergency box, double the size of the green case. Inside were enough supplies to keep Will, Stefan and myself alive in a freezing blizzard for up to two weeks. Tents, polar sleeping bags, dry clothing, equipment to make a fire and two weeks' worth of freeze-dried food were just some of the items. It was so well equipped there was even a toilet roll sealed in a clear plastic bag! Before leaving the station, as we did every day, we had entered our intended plan into the station's computer system, detailing the time we left, where we'd gone and our expected time of return. This was compulsory for anyone leaving the station and in the event people didn't sign back in on time, an alarm sounded, alerting staff. The last thing we wanted was to have to use any of the emergency kit and as wind kept forcing my skidoo off course I wondered how on earth we'd manage to pitch a tent!

After almost two hours inching our way along the track, my GPS told me we'd arrived, but looking ahead of me I couldn't see a thing. Visibility hadn't improved at all but all three of us were still together. I slowly brought my skidoo to a halt and looked around. Complete darkness was almost upon us and I felt as if I was surrounded by enormous grey walls. The wind still pummelled us from the east and the left side of my clothing was coated in a thick layer of snow. I knew the station must be just metres away, towering over my skidoo, yet I couldn't see it. I edged a little further. According to my GPS, I was there. I started to doubt it. Had I really put my life in the hands of this little electronic plastic box? A sense of panic washed over me. Was I standing next to Neumayer's front door or was I ten miles in the other direction? Not being able to find it, I felt that freezing to death just metres from safety was a real possibility. All of a sudden, like hitting a kerb, my skidoo bumped to a stop. In front of me was one of the two snowcats parked at the front door. I'd driven straight into it. We were back!

Slightly shaken, we hung up our kit to dry and warmed ourselves up. Walking down the corridor to make myself a coffee, I passed Dr Tim. 'Mr Lindsay,' he said in his usual happy German accent. He had no idea what we had just been through!

That evening, we spoke about our experience, both Will and I with wounds to show for it. Will's fingers had blistered back up and each tip was numb as he inspected each one carefully. Between my eyes, a large red raw patch had already started to scab over. Fortunately, it wasn't quite frostbite, but almost. Its precursor, frost nip, had taken hold across the bridge of my nose. The hours sitting with my face into the wind and the skidoo journey without any goggles had taken their toll. Dr Tim checked our wounds over and gave us the all-clear. 'They'll heal,' he told us.

In the living room sitting alone on a sofa I thought about the penguins' behaviour I'd witnessed and managed to capture on camera, and about how the three of us had picked our time to leave the birds and begin our journey back to safety. In hindsight, we had probably stayed a little too long, but without the guidance from both Will and Stefan I felt I would have pushed it too far.

I always have had a tendency to push too hard. I'd once returned from a filming trip with shingles, I was so exhausted. Was this another one of those occasions? As I sipped my coffee I contemplated telling Becky how close we had come. She wasn't aware of the extent of the danger I was putting myself under, even on the pleasant days. I decided it was probably best left that way; she had enough to be worrying about.

As we approached 21 June, midwinter's day, we'd been without the sun for over a month. Back in the summer, with my first ever exposure to twenty-four hours of daylight, I'd not experienced any problems. But being *without* the sun for twenty-four hours a day I'd started to feel some side effects. Mentally I felt OK, but physically it began proving extremely hard work.

Sleeping became impossible. Returning from filming days when the cold was sapping every ounce of energy, all I wanted to do was curl up in bed, fall asleep and recover. But it was hopeless! I went to bed consistently between 7pm and 8pm every night but by 2am I'd still be lying wide awake. I found it incredibly frustrating and the feeling of anger didn't help. Lying in the dark with my eyes wide open looking towards the ceiling of my room, I wondered what I could do. I'd always been a good sleeper and had never had any issues; early to bed, early up, my whole life.

With the lack of sunlight, Dr Tim had offered me vitamin D supplements in the form of tablets but being my stubborn northern self I thought I could get by without. I regrettably told Becky over the phone one evening; she had no sympathy, having to wake up and feed a newborn baby every two hours! I just couldn't understand it. The gymnasium hadn't seen me for a few weeks and as much I wanted to ride the bike I just didn't feel I had the energy. The only strength I had was used up each day out on the ice and every evening or day off I had nothing left to give.

With no stamina to do anything physical, when not filming, boredom quickly began to set in. I dived into my two aluminium personal boxes that sat in the corridor just outside my bedroom door. It had been so long since they'd arrived, I'd forgotten what I'd packed in them. Looking for items to fill my spare time, I wanted something that didn't involve raising my heart rate to 190 beats per minute! The violin remained untouched, the unicycle was still in bits but at the bottom, two large plastic boxes of fly-tying materials and tools looked inviting.

As a child, I'd grown up fly fishing. My uncle and late granddad had been keen anglers and had spurred an interest that had never left me. Being out on the riverbank I'm sure had a huge impact on me and my love of nature. Fly fishing the remote Cumbrian tarns and rivers for salmon and small wild brown trout was my favourite way to relax. I'd see otters, fish jumping and on the high fells, ring ouzels singing; it wasn't just about catching fish. I'd also learnt a great deal. Fishing with artificial flies had forced me to recognise what insects or small fry the fish were feeding on and by replicating that on the end of my line I gave myself a chance of catching one. Fishing was a way of getting in tune with nature and without doubt helped me develop the patience required to film wildlife.

Using cotton thread and old birds' feathers, I tied materials round small hooks in various patterns to imitate different forms of life. I hadn't touched any of my fly-tying kit for at least ten years and when packing my boxes back at home to send to Antarctica I thought the long storms, during which I knew I'd be confined indoors, would be the perfect excuse to rediscover the art. I dragged the spare bed out of my room and replaced it with two small tables for my equipment. A tall lamp that I'd stolen from the corner of the living room brightly lit my workspace. With my materials spread out I fixed a small hook into the vice and started wrapping thread around it. Immediately my head space changed and for the first time in weeks I felt relaxed. I hadn't been stressed, just exhausted, and tying flies began to occupy my mind.

Across the world, fly-tying is an art in itself and even though I'd tied flies as a teenager with the aim of using them to catch fish, for many fly-tiers it was just a hobby. I'd always struggled to understand why people tied flies that they would never use to catch fish, but as each evening fell and I sat down to tie the same pattern over and over again, I began to understand. My room was silent, the only source of light was the lamp that shone over my vice and as I sat alone, my mind began to drift. I felt as if I was sitting on a rock beside the river at home, rod in hand while kingfishers bolted past like blue bullets. I tried to envisage how the fly I was tying would look in the water, rising and dropping through the currents. To my eyes it looked attractive but would fish think the same?

My uncle had bought me my first fly-tying kit when I was a young boy and thinking back I remembered one of my first flies. It was a small hook with a shiny gold bead on its head. I'd sneakily snipped a bit of cream hair off my cat's tail while it was sleeping and tied it into the end of the fly. The day after, it caught me a trout and the feeling of achievement that gave me as a child was unparalleled. I'd never felt anything like it. Keen for Walter to one day experience the same, I started working on tying him a selection. With no cat to have a go at, I wondered how creative I'd have to be.

For three days, the blizzard outside hadn't shown any signs of subsiding and for the first time I began to feel sorry for the male penguins out there having to endure it. I hated not being able to see how they were coping. Fly-tying was taking up a lot of my spare time and to stretch my legs every now and then I took trips to the kitchen to refill my teacup.

Although we were sheltered and warm inside, I could still feel the effects the storm was having as it battered the outside of the station. Wind speeds outside hadn't dropped below eighty miles per hour and the long east side of the station was taking the full brunt. In the storeroom opposite the kitchen, two metal rack units, each with shelves running the length of the room, were piled from front to back with bottles, jars and cans. Anything that didn't need refrigerating was in there. Bags of flour, jars of jam and marmalade, pasta sauces, olives and coffee beans. As the storm outside had intensified, the station had begun to shake. Vigorously vibrating glass bottles of olive oil and vinegar rattled against each other as they wriggled closer to the edge. Air-conditioning units that vented into every room clattered as if plastic plant pots were being thrown down their shafts.

With our skidoos still parked outside, covered in their flimsy sheets of fabric, I went down the central staircase to check they were still where we'd left them. Looking out of the window, I saw the wind was swirling tornadoes of snow through the air. Despite being less than a couple of metres from the door, all three skidoos and the two snowcats were invisible. I shot back upstairs and put on my thick red polar suit, a hat and a pair of goggles and asked Will to accompany me back down. Even though I was only stepping outside, from our eventful journey back from the colony I'd learnt it was an incredibly dangerous thing to do. Feeling as if I was opening a space airlock, I opened the door that led from the bottom of the stairs and closed it behind me. Now standing in the small porch with a door to the outside in front of me, I pulled my goggles over my eyes. A thick layer of snow coated the walls, having been forced through tiny gaps under the door by the wind's force. Kicking a pile of snow away from the base of the door, I unlocked the handle. Like smoke from a fire, a plume of snow immediately filled the room. Taking one step outside, I was promptly taken off my feet and blown across the ice into the parked snowcat. I lifted my head. Despite my having been blown just a few feet, the door had disappeared. Snow began building up on the inside of my goggles. Feeling my way around the back of the snowcat on my hands and knees, I reached the skidoos. Although they had the protection of a large vehicle parked either side, the covers had been blown off and were anchored down the side of each skidoo by heavy banks of frozen snow that had built up on top of the fabric. All the plastic cable ties had snapped on my carefully modified covers. But in such violent conditions there was simply nothing I could do.

Deciding to abort, I dropped back down onto my knees and began to make my way back to the door. The crackle of wind and ice hammering against my clothing was deafening and, inch by inch, I pulled myself closer, or so I thought. After a few minutes of blind crawling, my hand hit metal. I was nowhere near the door! Lying flat commando-style to keep below the wind, I pulled myself along the ground wondering if I was pulling myself towards or away from safety. All of a sudden, I felt a tug on my hood. Will, who was standing in the doorway, grabbed hold and pulled me through. As I leant propped up against the wall Will slammed the door shut and returned its long handle to the locked position. It felt as if we were in a horror movie preventing a persistent monster from entering. I'd never felt so scared and despite my having travelled literally only a few metres, the ferocity of Antarctica couldn't have been more obvious. How on earth were the penguins surviving?

Eventually, after day six, the blizzard began to subside. Wind speeds had rarely eased and I was desperate to survey the damage down amongst the penguins. The severe temperatures over the previous few months had taught me the need to properly prepare myself. It was time to get kitted up and get out there. A standard layer of thick thermals tucked into heavy woollen socks was first. Chunky fleece-lined wading pants I used for fishing back home slid over the top, while a T-shirt, jumper and thin gilet encased my chest. The heavy fleece-lined salopettes that I'd been provided with were pulled over the top before my enormous red polar suit finalised the look. A balaclava and two snoods tucked into my jumper around my neck covered my mouth and nose, and finally I donned a blue bomber hat with large flaps that strapped around my chin and a gigantic pair of orange snow boots that laced up around my ankles. I'd mastered the art of getting ready each morning, doing everything in a particular order, but even so it took around twenty minutes and some days I felt exhausted just getting dressed. Fully kitted up, inside I'd overheat within minutes, so stepping out of the door into −50 was actually a huge relief. After the fully air-conditioned station, my lungs revelled in filling with fresh, clean air.

Ready to depart, I looked over to our skidoos. Only the tow hitches of each one remained visible. Buried under four feet of snow, the three lined up next to each other formed the foundations of a hill of ice. We had detached the sledges to strap down our skidoo covers, and all three had been blown over fifty metres away across the ice. It was problem after problem and I began to think we weren't going to get to the penguins at all. With three long-handled shovels, Will, Stefan and I began digging. After almost half an hour shovelling away the last of the snow, we discovered another issue: the engines were packed solid. Small holes in the air vents had allowed particles to filter in and every gap had been filled. We were getting no response from the jam-packed engines when trying to start them, so we had to remove the plastic casing and try to clear out as much snow as possible. As it had solid-ified into an enormous chunk of ice gripping every pipe and cable around the metalwork, this was painstaking and deli-cate work.

Eventually all the engines were clear and the sound of three skidoos blared out. Steam rose from the air vents as the snowy residue melted away under the engines' heat. The track had disappeared under heavy drifts of snow and negotiating the mounds along the line of flags was slow. As I approached the edge of the elevated shelf to look down onto the colony, I saw long lines of penguins leading into the distance. It felt nice seeing them again but I was desperate to see how they'd fared. Over the course of the storm the prevailing winds had pushed the rotating group of penguins almost a mile across the sea ice and with nowhere left to go other than tight up against the cliff of ice, they'd begun to march in single file back to where they'd started. One by one, each of the 5,000 male birds peeled away from their huddle and, still carrying their precious eggs, shuffled across the ice. The storm had left them battered and exhausted but with more bad weather inevitable, there was no rest.

As I looked down onto the birds as they relocated I wondered what they knew that I didn't. Why were they bothering to move back to where they had started when it didn't seem necessary? The area of ice they were on now appeared just the same as that they'd spent all autumn and winter standing on, and with a cliff either side of them in the small inlet they'd been forced into, it appeared more sheltered. But having spent so much time with the emperors, I knew they weren't daft. As with everything in nature, there must have been a reason, but what was it? Did they feel the ice was less stable? Were they worried the females wouldn't find them in their new place? All sorts of possibilities went through my mind.

As the whole colony began to relocate to where they'd started I moved my skidoo further down the ramp and onto the sea ice. Across the landscape were dropped eggs, each with a single straight crack from top to bottom. The liquid embryo inside had frozen and expanded, forcing the shell to part slightly. Drifting snow rose and fell over each egg with the wind. They'd become part of the landscape. It was my first encounter with death amongst the penguins but surprisingly, it didn't affect me in the way I'd thought it would. Seeing the odd bird walking around without an egg, however, did. Knowing the enormous effort they'd put in to caring for that egg, I couldn't help but feel for the males. The females who were feeding at sea had no idea. Fattening up and collecting food to return with to feed their chicks, they weren't aware that some of them were now doing so for no reason. Even though eggs had fallen and a few of the males had failed in their bid to produce a new generation, I was amazed I couldn't see more. I couldn't believe the viciousness of the blizzard hadn't generated more casualties and my admiration for the male emperors grew.

As I approached the shifting colony, directly ahead of my skidoo lay a black mound on the snow. With the sea ice so pure and white, anything positioned on the snow was incredibly obvious. I drove over and stopped before I reached it. Jumping off, I immediately knew what I was approaching. It was an adult emperor lying dead in the snow. Despite having succumbed to the conditions, it was unmarked and its plumage appeared just as regal and colourful as that of the active birds behind. The only suggestion the emperor wasn't alive were its eyes. Once so perfectly black and shiny, they'd frozen solid and formed a crispy layer. I laid my mitten on its flank and even though its body was frozen hard I could feel the pure muscle and power beneath its carpet of pure white feathers. Between its outstretched feet and one of its wings, the tip of an egg protruded through the snow. Before me were two lives that had fallen victim to the storm. Yet to reach the middle of winter the male emperors still had a further month to go until their chicks hatched and their partners returned. The average lifespan of an emperor penguin is twenty years and I wondered how old the bird that hadn't survived was. Was this its first breeding year or its fifteenth? I looked up at the line of birds that were beginning to re-form their huddle and wondered what more they'd have to go through.

Why they were relocating continued to baffle me. The colony had always been on the move but in a circular motion, drifting away from and around stained areas of ice that they'd contaminated with their guano. So much snow having now fallen to cover those areas, that didn't appear to be the reason, and I wondered if I'd ever find out. It was certainly behaviour that hadn't been documented before and with temperatures tipping −50 once again, the birds were quick to resume their huddle, packing tightly into one single group and again falling silent.

Just like back home on frosty clear mornings, the cloudless conditions were proving the most painful. Overcast days with a layer of insulating gloom sitting above Atka Bay felt considerably milder and I found myself referring to anything above −30 as 'warm'.

With the emperors settled and not displaying much behaviour other than huddling, we were freed up to explore the icebergs I'd so longingly wished for over the summer. About six kilometres into the central part of the bay a small cluster had frozen in place. I'd looked over at them in the distance from the penguin colony for months, wondering what they actually looked like close up, but with our time filled by having to document key moments of behaviour, we'd been unable to inspect them.

We had not travelled that far from the station before, so we decided to try to get to them only on settled days. The tides that continued to rise and fall on the ocean had the potential to destabilise the ice surrounding the icebergs so, as always, we had to take extra care. Glancing back, Neumayer looked within easy reach across the flat, featureless ice as it did on every clear day, but as we drove behind the first berg, it disappeared. At over fifty metres in height, three pinnacles of ice towered over the bay.

The icebergs, which had once formed part of the continent, were blocks of frozen fresh water, rather than saline liquid from the ocean. Whether they'd been calved off an ice shelf or a glacier further around Antarctica was discernible from their shape. Large tabular icebergs with flat surfaces were characteristic of having come from floating ice shelves, just like the one that had broken away from the North-Eastern Pier, whereas smaller, more irregular icebergs could have broken away from more jagged, steeper glaciers that fell directly into the ocean off the continental rock. Deep blue cracks and seemingly never-ending caves led into one iceberg's centre. With its flat surface lying at a steep angle, the iceberg at some point in its life had split into two, its heavier side tipping over revealing the rounded, beautifully smooth underside that had lived its life under the water. Up to ninety per cent of the mass of an iceberg lies under the surface and it surprised me looking at its height just how deep the ocean below my feet must be.

Rather than having hard edges that erupted out of the snow, a ramp led up from the sea ice onto the iceberg's curved base.

The light-blue ice felt solid, as hard as rock, and the ripples that had formed by the saltwater eroding it made it look like blown sand on a beach. With so many bergs peppered across the area of ice, we began naming them. The one with large angular arms that dominated its appearance I named 'Yosemite', after the American national park famed for its rock formations. Another iceberg covered in deep caves we called 'Flintstones'. A single tall spike of blue ice that resembled a fresh dollop of ice cream became 'Mr Whippy'. We'd been so fortunate in what the ocean's currents had delivered to us over the summer and the filming possibilities were endless. Along with the sastrugi snow patterns that had formed across the surface of the sea ice, beauty was everywhere. As I stood and admired the marvellous structures it occurred to me that what I was looking at was simply water. The entire landscape on which the penguins bred and on which the station sat was just frozen water. It was such a complicated landscape, yet in reality so simple.

After 188 days away from home, midwinter was upon us. The term 'the shortest day' had no relevance in a part of the planet where the sun didn't rise, but as was tradition across the frozen south, we planned to celebrate with a day off. Reaching such a milestone unharmed both physically and mentally felt like a huge achievement considering the environment we were living in. The twelve of us at Neumayer had got on incredibly well up to this point, rarely having any disagreements, and I felt it was important to share such a unique occasion properly. For me it was certainly the only midwinter in Antarctica I'd ever experience.

Just as he had in December, Will dived into the storeroom

and decked out the whole living room and dining area with tinsel, baubles and a small decorative Christmas tree. Despite being the middle of June it felt incredibly festive looking out of the window onto the dark-blue expanse of ice and snow. He even projected a looping clip of a roaring log fire onto the drop-down screen in front of the sofas.

As was traditional amongst Antarctic research stations, electronic greetings cards were emailed around the continent wishing fellow overwinterers peace and health for the remainder of their stay. Messages began appearing on the noticeboard on A4 sheets of paper as Dr Tim printed out each one he received. Photographs and messages from overwintering stations including Palmer Station (an American base), Jang Bogo Station (South Korean) and Casey (an Australian base) demonstrated the variety of nationalities represented on the same continent. Only approximately 1,000 people spend the winter on Antarctica, an area that is roughly the same size as Europe.

Personal traditions at Neumayer included opening a large wooden box sealed tight with screws. Having been brought down on *Polarstern*, the large box was full of bottles of wine donated from a vineyard back in Germany as a gift. Over the large goose that our chef had roasted especially for the day, we sat down and opened a couple of bottles of red and white, each congratulating one another and wishing for our final few months on our own to be safe and happy. I dug out a fancy dress costume I'd brought with me and surprised everyone at the dinner table. A pair of trousers that made me appear to be sitting on the shoulders of a snowman was apt for an area covered in the white stuff and everybody found it very entertaining. It was a special afternoon that I will remember forever.

After dinner, I grabbed a handful of orange balls and a couple of golf clubs and challenged Will to a game outside. The day itself had been one of the few calm days on which we'd not filmed and it seemed a shame to waste it by not going outside, even if the temperatures screamed otherwise. Climbing a hill at the back of the station I teed up, scraping a tiny mound of ice using the blade of my seven-iron. Sitting the ball proud, I prepared to swing.

'Around the station and first to hit the skidoo tent!' Will challenged. I fired my orange ball high into the sky and gravity took hold of it, powering it hard onto the surface of the snow. Like hitting rock, the ball rebounded up into the air, not leaving a single mark on the solid ice. I worried for my clubs. They certainly hadn't put up with these conditions before. Will played his shot, and the same thing happened. With the worry of putting a ball through a station window (which we didn't think would go down well) we picked them up and positioned them closer to the skidoos, further away from the station. A single shot later Will had won the coldest, quickest and most remote game of golf we'd ever play.

Later that evening I received a message from Becky. She was desperate for someone to talk to. I had been in Antarctica exactly six months. I knew how she felt. The long stormy days when I wasn't able to do much had started to get to me and on occasions I found myself ringing her in tears just wanting to be home. Over the space of a couple of days I was going from unthinkable highs to incredible lows and at times I just wanted to get out.

On the monitor that displayed the outside temperature and wind speed was a section that flicked between various webcam images from around the world. Our IT technician at Neumayer had managed to add a webcam looking down my local river back home. Every time the picture appeared I couldn't help but stand next to the screen and study the image for a few minutes, admiring the beautifully clear water, the fresh green leaves on the trees and the warm sunshine that the river bathed in. I'd grown up swimming under the bridge and watching the salmon and sea trout swim by. I'd see cyclists crossing and at times all I wanted was to go home. Again and again, I had to remind myself that my Antarctic stay was only temporary and I knew deep down that the last few months would be over in a flash. Sometimes I forgot how lucky I was being there and how many people would have given anything to be in the same position.

To divert my mind I started thinking of things I could do for Walter while being so far away. Other than images and short videos of my face on Becky's phone, I needed something for Walter to associate with his dad so that when we finally met it wouldn't be such a huge surprise. Draping a spare duvet from the bunk above my bed and filling each end with pillows, I created a makeshift sound studio to record some stories for Walter. Night after night, I read out the twenty-three children's tales by Beatrix Potter into my microphone, starting with *The Tale of Peter Rabbit*. I'd grown up enchanted by her stories and the fact that she had lived locally to where I had been brought up in the Lake District meant I'd been fascinated by her rural life and work. I remember reciting *The Tale of Peter Rabbit* word for word on my journey to school every morning as a child. The stories had brought my imagination to life and I felt that if I could pass on the same magical feelings to Walter, the natural world would become a big part of his life.

6.

New Life on the Ice

As we entered August, now deep into winter, I had been in Antarctica for eight months. Although I was well accustomed to life at Neumayer, I was still feeling the effects of the previous month, which had lowered my morale and physically pushed me to my limits. I was well over halfway through and every day that passed I was counting down the days to meeting my little boy. Receiving photos and videos every day, I felt like I knew him almost as well as Becky did.

According to calculations the sun had returned on 22 July but due to persistent heavy cloud I'd not seen its golden glow since it had set well over two months before. Even though it had been rising above the horizon for lengthening periods each day, it brought no heat to the frozen landscape. My spirits were still low at this point; I was desperate to see the sun again to give me a bit of a psychological boost. I also clung on to the hope that my sleep and health would improve with its return.

A huge weather system had blown its way in from the east and Neumayer had been shaking from the battering it was taking from the winds for over a week.

In my diary on the page dedicated to 27 July, my sister's birthday, large letters read 'Chicks – due from today'. The sixty-four-day incubation period was up and from this point I was expecting new life amongst the huddle of males. Since the last of the petrels and terns had flown north to escape the decreasing temperatures of autumn, Atka Bay had been home to nothing but the emperors and a few Weddell seals. New life was an enormously exciting prospect and after such a physically difficult month it was something I was desperate for. I was relying on the arrival of the chicks to help lift me.

The storm that had prevented me from even attempting to go outside had been the strongest all winter and there wasn't anything I could do other than stare through the window by the front door. The nearest of the sixteen stilts holding the station above the snow stood only a few metres from the window, but only briefly did it appear from behind the horrendous waves of ice-filled air that blasted horizontally through under the station. The prospect of attempting to stand outside against its full force was terrifying and made me seriously worry about the penguins. What they were having to endure was beyond my imagination. I had no idea that weather on Earth could actually get this bad, let alone that there was a creature that could survive in it.

Plus, the incubation period was over. Some of the eggs would be hatching any time. The thought of newly hatched chicks being born into such brutal conditions was harrowing.

Surely they wouldn't survive if they hatched now? Having filmed the colony of males in a storm the previous month, I thought I had an idea of what they were going through, but this was so much worse than what I had experienced with them. With a week of relentless freezing cold winds with not an ounce of respite, I could only imagine how exhausted they would be. Wind speeds were well over double what I'd filmed them in and even then it had appeared that they were fighting for their lives.

Part of me was desperate to be with them, not only to witness what they were doing but also to ensure that they were all actually still alive. Although I was desperate to see the penguins, visibility was so extraordinarily bad that I would have had to be so close, almost part of their huddle, to have actually seen them! Even though I worried about the emperors, the fact that they could survive conditions in which humans were unable to observe them amazed me. Conditions were tough, but this is what the emperors were adapted for.

It wasn't just the males that were battling. At least thirty kilometres north, where the sea ice turned to loose pack ice, the females were feeding, or at least that's where I presumed they were. Diving deep under the ice to feed in such rough conditions must have been tough. Extremely little is known about what the females get up to during an Antarctic winter. They spend their time in places so inaccessible to humans it's basically impossible to follow them. I wondered how they were coping. Were they also huddling? Did they need to rest on ice floes? How were they keeping warm? Thinking about the females' survival out near open water bewildered me even more.

Sitting inside at this time was difficult. There was a chance that chicks were beginning to hatch in the colony, which made the situation hard for me to come to terms with. This was the moment we'd all been waiting for and absolutely everything the birds had done day after day up to this point was for this hugely significant moment. Unable to do anything, I stood for half an hour with my head leaning against the window. My thin woolly hat was the only thing that separated the skin on my forehead from the cold glass. Even though the sun had returned, the day length was still short, giving a potential maximum of six hours' filming on a clear day, but with the weather like this I couldn't even manage one. All I could do was sit and wait and hope that some of the chicks would cope with such a brutal beginning to their little lives.

It had only taken me the first couple of days of the blizzard to rearrange all my camera kit. Everything was ready to go for when it finally relinquished its grip, but with the winds continuing I had time on my hands. No filming meant that Will had finally caught up with processing the mountains of footage I'd been returning with each day and was able to have some of his own time off. Will had much more physical energy than me throughout polar night and without his positivity through this difficult period I'd have struggled. He'd always been a keen runner and each morning before beginning work he had started training on the treadmill with the aim of completing a marathon outside while in Antarctica. Having already completed the London and Berlin marathons, Will was on his way to achieving the holy grail of the 'Big 6'. All he needed to complete were the Tokyo, Boston, New York and Chicago events. His idea of creating the 'Big 7' and adding an Antarctic marathon to that list was bonkers.

Before we'd arrived, we'd both thought we would have lots of downtime and really wanted to get super fit while in Antarctica, but as it turned out we didn't have as much spare time as expected and at times we were really lacking in energy. Trying to find the motivation to go on the exercise bike was difficult for me, so instead I put the unicycle together. I used the pressurised air machine in the workshop to pump up the tyre and repaired a puncture with a spare inner tube I had brought with me. Unbelievably, I hadn't been the only one to bring a unicycle to Antarctica. Our chef was also desperate to learn how to ride one and had brought one down to learn on. As a small gift over winter the mechanic had erected a mobile scaffolding tower just outside the generator room incorporating a long horizontal handle for us to hold onto while practising.

For hours I sat alone rocking backwards and forwards trying to maintain my balance. It didn't take me long to be able to release my grip and let go but I found it impossible to move forwards and backwards. I'd always considered myself very lucky that I'd been able to pick up new skills and sports quickly. As a kid I could turn my hand to anything, so not being able to make progress on my unicycle was incredibly frustrating. My ankles were black and blue from where they'd been hitting the cranks and my thigh muscles were constantly aching from tensing up to maintain my position on the seat. Although I was desperate to learn I came very close to giving up. My patience was definitely wearing thin. I needed to get outside to reactivate my brain as soon as possible.

As the storm continued relentlessly, I wondered if it would ever give in. Had eggs begun to hatch? What was I missing this time? One evening in the living room as a few of us watched a film on the large screen, I momentarily switched my focus to the computer screen displaying the outside weather conditions. Watching the wind and temperature fluctuate had become an obsession as I'd made every effort to be with the penguins when the conditions allowed. Halfway through the film I glanced over at it. The wind was hovering around ninety kilometres per hour with a temperature of −40 including wind chill and had been like that for days, showing no signs of easing off. A few minutes later I checked again. The wind was now five kilometres per hour and the temperature was −4! Within less than five minutes the wind had dropped to a level I'd not experienced for a month and the temperature was that of summer.

In my shorts, T-shirt and slippers I ran downstairs and opened the front door. Lumps of soft snow fell from the top of the door frame. A floodlight lit the platform under the station and as I stepped out the air felt warm. I was baffled at how quickly the weather had changed. I'd experienced storms coming in quickly but not passing over as rapidly as this. Running back upstairs to check the forecast, I went straight back to the computer in the living room. No sooner had I got back to it than the wind speed returned to ninety kilometres per hour! Looking at some live data of Atka Bay I could see the station was right on the edge of the enormous weather system that had enclosed us for days. As it had briefly swung slightly to the north and with the protection of an insulated dense layer of cloud, conditions had altered rapidly. From what I could see

I doubted the penguins had been lucky enough to experience the same lull in the storm; they were only a few kilometres north of the station but not close enough to have experienced the short break in the winds.

Following more than ten days of brutal, relentless wind, the weather finally cleared overnight and we were free to make our way back down to the colony. Yet again the first half an hour was spent digging out the skidoos, this time with the help of some of the other guys. All three machines were completely submerged under snow. A large snowcat parked either side of our three skidoos usually acted as a bit of protection from the wind but this time the snow had completely filled the gap. The snow was dry and soft and each heaped shovelful weighed next to nothing. It felt as if each shovelful was empty. Although the remains of the track down to the birds appeared non-existent, amazingly the cane flags still marked every hundred metres of the six-kilometre route. The edges of the flags' fabric had become frayed but unbelievably all but a couple of the tall canes stood undamaged. The force of the storm had snapped just two and it impressed me what they'd managed to withstand. Despite the track being covered over with soft fresh snow, the solid base underneath, which we'd driven up and down and flattened so many times, remained. As the skis of my skidoo glided through the loose mounds it felt as if I was driving on a smooth, newly ploughed path.

Within ten minutes I could see the eight flags that marked out our ramp down onto the sea, but to get my first view of the colony in nearly two weeks, I diverted off the track towards an elevated edge of the shelf; both Stefan and Will followed. With such an extended period of strong winds having prevented me from observing the colony, I'd expected the birds to have been pushed right up to the base of the ice cliff, effectively to a dead end. As I slowed, both Will and Stefan's skidoos came up alongside mine, all three of us eager to get a glimpse of the colony. Despite having had only a few weeks above the horizon since its two-month absence, the sun was already sitting bright and high above us. The wind, however, hadn't completely eased off. Every now and then light gusts erupted, blowing up loose snow into small plumes, spiralling them into the air.

As we slowly came to a halt near the edge, the mass of penguins below began to appear from behind the ridge. The moment all three skidoos stopped, the ghostly figure of a snow petrel glided across in front of us. It was the first sign of life, other than the emperors and Weddell seals, that we had seen in nearly four months. As it turned its wings and circled around us just millimetres above the ice I wondered where it had been and how it had coped with a winter at sea. Had it been accompanying the female emperors fishing? Conditions must have been horrendous, yet its beautiful elegance showed no trace of what it had endured. Its white plumage didn't have a blemish on it.

It began to glide back along the shelf edge and I turned off my engine and returned my focus to the penguins. Will and Stefan did the same; all three of us sat on our skidoos overlooking the colony, which lay about a hundred metres away across the sea ice. They hadn't moved as far as I'd predicted and within an instant the sound of calling penguins hit me. Having been silent for the best part of three months the cacophony of serenading emperors had been restored. Having had no reason to make any noise over the dark depths of winter, what had changed all of a sudden?

I scrambled the camera with its long lens and surveyed the distant birds through it. Some were scattered individually around the periphery of the colony. Others, with their heads down, huddled against the cold. Even though the majority of the colony remained tightly packed, the area the colony occupied on the ice was larger than I'd seen since the birds had mated. I pulled down my balaclava, exposing my nose and mouth, and lifted up the flaps of my hat so I could hear the colony in more detail. I'd missed the sound of calling emperors.

As I looked north to try to catch one last glimpse of the snow petrel heading back in the direction of open water, my eyes were distracted. On the horizon, blurred against the face of a distant iceberg, were black specks. Highlighted against pure white, they stood out. There had been nothing in the north since the females had left at the end of May. At first I was confused. Had the storm wreaked so much havoc that hundreds of eggless male penguins had left the colony? Was the colony calling in celebration of the return of good weather? I placed my large mitt-covered hands up against the side of my head to funnel the sound directly towards my ears.

Chicks! Other than on the television, I'd never heard the sound of a newly hatched emperor chick, but its begging call was unmistakable. I shouted to Will and Stefan and listened again, the mass of trumpeting birds making it tricky to hear any other calling chicks. Nothing. I began to wonder if my ears and eyes were playing tricks on me and I had got this all wrong. Looking into my viewfinder, I scoured the nearest edge of the colony, using the camera as a substitute for my binoculars. Panning slowly from right to left, my eyes quickly assessed every bird I could see.

One bird grabbed my attention; he was leaning forward and concentrating on his brood pouch. I locked the head of the tripod in position and stared at his feet. Being a few hundred metres away the adult bird was small in my frame and I couldn't zoom in any closer. I watched intently. As other penguins shuffled by, he maintained his concentration. All of a sudden, a chick appeared, poking its head out from between its father's feet. It was my first distant view of a baby emperor penguin. Despite arriving during an incredibly challenging storm, chicks *had* been successfully hatching. I couldn't believe it. Weighing in at only 200–400 grams and with only a very thin layer of down, the chicks didn't have the capacity to keep themselves warm; they had done well to survive their first major challenge.

Desperate to get down and closer, we rushed the equipment back into its dedicated boxes. For ease, I held the camera on my seat between my legs rather than disassembling it and, using my spare hand, I drove down the ramp onto the flat sea ice. There were penguins everywhere so, keeping tight against the cliff of ice, we drove along the bottom edge of the shelf to a position just below where we'd been sitting and watching. Being closer to the birds I was able to look at each one in more detail. I couldn't see the individual I'd just had my camera focussed on but a lot more males were showing signs of being fidgety. My eyes didn't leave their feet as I searched for signs of new life amongst the thousands of birds that stood in front of me.

Being down at the same level as the penguins, I could barely hear myself think, the characteristic trumpeting from hundreds of calling individuals creating a deafening choir. Again, I was keen to film events in chronological order. Having seen one chick, I knew there must have been more, but it also meant somewhere there must be eggs in the process of hatching. Copulating had taken place over a two-week period and possibly for a week or two before that, when I'd not had access on to the sea ice. So long as the weather held I had at least another ten days of the hatching process in which to capture the magical moment of a chick breaking its way out of the egg. It was potentially one of the most intimate occasions during the emperors' year and I simply had to record it.

I continued to monitor each agitated male bird in the hope they'd give me a glimpse of their precious egg. As they had been throughout the winter, they were very reluctant to expose their treasure, especially now that they had managed to deliver it unharmed to such a vital stage of the process. The birds were more active than I'd seen for a while. The improvements in the weather allowed them to stand apart from each other, which meant gaps allowed me to focus on birds deeper into the group.

I clocked an emperor between two others behaving in a similar manner to the one I'd seen from the top of the cliff. I watched him in detail in the hope I'd get a glimpse of his egg. Constantly preening the outer edge of his brood pouch, he appeared extremely unsettled. None of the males had given their feathers this much attention since just before receiving the egg from the female back in May. Running the end of his bill from the base of each feather along to the tip, he arranged each one, making sure they were in perfect condition. Despite the white feathers around his feet being so close to the ground (and the backsides of other penguins) they appeared pristine and whiter than the snow. It amazed me how clean he'd managed to keep them without entering any water to clean them off.

With no warning he rapidly uncurled his plumage. With his egg exposed, resting on his scaly feet, he gently rolled it around, readjusting its position using the tip of his beak. One by one, his long white feathers folded back down, hiding the egg again. Immediately the emperor lifted his pouch once more. I zoomed right in to grab a close-up shot, the mango-sized egg filling my viewfinder. Right in the centre was a long crack and a small hole, with the tip of a chick's beak poking out. As it chipped its way around the inside of the egg to break out, it chirped through the hole it had made. I could see its tongue as it made its father aware it was trying to break free. Falling into frame, the male emperor's bill gently made contact for the first time with his new chick. It was a magical moment and one that I will treasure forever.

For over two hours I patiently sat watching and waiting for my opportunity to record the chick's emergence. This was my chance and I didn't want to miss it. Each time the male revealed his egg, the crack had deepened and the shell was a little more open. The tiny chick, who had been chipping away at the inside of the shell for the previous twenty-four hours, was finally able to lift the lid and prise itself out. With large parts of shell falling away onto the ice and the soft membrane flapping in the light breeze, the young and weak emperor uncurled its folded neck. Covered in a damp, delicate layer of light-grey downy feathers, its skin showed through, making it appear naked.

Covered in ice, a fluffy emperor chick enjoys the
protection of one of its parents.

During polar night the bright moon and stars illuminate the huddle.

In temperatures around -50 degrees Celsius, individuals squeeze into every space available to huddle for warmth.

The moment everyone had been waiting for:
an emperor chick begins to hatch.

Two proud parents show off their precious chicks to one another.

Females with bellies full of food appear out of the blizzard following sixty days out at sea feeding.

A chick that succumbed to the weather lies half buried.

An aurora australis lights up the sky above the station.

The beauty, silence and grandeur of 'Yosemite' iceberg.

Filming from the ice shelf, snow ramps leading down on to the sea ice.
(Photo by Stefan Christmann)

Ice particles in the air form 'sun dogs' and a halo behind the colony.

Independent chicks rely on each other whilst their
parents are out at sea feeding.

The wind whips up drifting snow against the shelf and into the deadly gullies.

Having returned following the brutality of winter, a pristine snow petrel glides along the top of the ice cliff.

A Weddell seal glances up at me as it enjoys the calm weather conditions that spring delivered.

Almost immediately the chick mustered enough strength to lift its head above its body and instinctively begged to its father for food. As he had not eaten for well over a hundred days, I was unsure he would be able to offer his chick anything. He'd fought numerous horrendous ten-day-long storms, moved on two occasions across the sea ice to a safer location and had been forced to his physical limits to maintain his own body heat to survive. His diminishing energy levels couldn't have supported him for much longer. Yet, standing upright, he appeared to be trying to muster something for his chick. He elongated his neck upwards, throwing his beak forwards. His neck went into spasm, forcing his wings to repeatedly lift up and out away from his body and back down. It reminded me of a child's toy.

Bending over, he gently offered his open beak to his chick, triggering an impulse in the chick to open its beak wide. Lifting itself as high as possible, the chick stuck its entire head into its father's mouth. From behind his spiny tongue the male regurgitated a thick creamy substance. I couldn't believe what I was seeing. Had the male really managed to store away food just for this moment? Surely after such a long period the food would be off? It fell down into the chick's mouth, its weight and gravity taking the chick's head with it. It was an almost comical sight. Half was swallowed, half fell onto the ice, immediately freezing and becoming encased in a silver crust.

Having received its first feed, the chick rested its head on its father's feet as he covered it with his feathers. I sat up. I had been leaning over with my eye glued to the camera and my neck was as stiff as the chick's must have felt. I watched as the parent settled down again, the chick disappearing under his heavy coat of feathers. The father closed his eyes, knowing he had done all the hard work of bringing the next generation into the world.

Looking over the colony at the distant birds that I'd seen from the elevated clifftop, it suddenly dawned on me. I grabbed my binoculars and looked across the horizon. Tens then hundreds of emperors appeared. The females were returning. Timed to incredible perfection, fat, healthy, colourful females were on their way back to the colony after their sixty or so days at sea. How, just how, had they managed to return at exactly the same time as the chicks were hatching? Unlike when the entire colony had marched in great long lines across the fresh sea ice, the females were tobogganing on their bellies individually. Rather than travelling in long lines, they were trickling back towards the group of males, with birds arriving into the colony every few minutes or so. Some arrived two or three at a time, others on their own, but looking out towards open sea there was a constant flow of penguins.

As they propelled their bellies across the fluctuations in the snow caused by the patterns of sastrugi, their bodies disappeared and reappeared behind ridges. I could see the fat they had put on over the winter ripple down their flanks as they powered ever closer. With their heads up above the snow, calling as they approached, a cloud of breath evaporated in the air as they slid underneath it. They were travelling at an incredible rate towards the colony. I changed position to try to capture a female returning back into the main group. I was desperate to see her reaction when she caught the first glimpse of her newly hatched chick.

Fifty or so yards before sliding into the group, a female stood up. Using her beak as a lever and her two wings for balance, she lifted her upper body effortlessly. Immediately, she called and listened. Impatiently, she continued walking on into the colony, stopping, calling and again listening. She seemed in a rush. She knew what she was returning to. With so much behaviour happening with the penguins, it felt fantastic to be with them. Witnessing arguably the cutest, most well-known creature on Earth emerge from its egg was a real honour, but with the return of the females happening at the same time it was extremely confusing and I found it difficult to focus.

The colony had suddenly ramped up into a hive of activity. For a couple of months I'd become accustomed to the silent huddle of males all trying to save energy in one group. With females returning and calling in search of their partners, males calling in celebration of their newly hatched young and actual chicks shouting for food, the dynamics had rapidly changed. Despite the dire temperatures of the last couple of months, on clear, still days the penguin colony had been a serene and peaceful place to be. With so much happening now, I felt the need to film everything at once.

Although I felt that my normal strategy of recording events chronologically would allow me to capture them in more detail, there was a niggling doubt in my mind, which was causing me mild panic. What if there was another two-week storm that prevented me from accessing the colony and made it impossible to record the hatching in the detail I wanted? The pressure felt overwhelming. Miles, our producer, was fantastic with his phrase, 'You can only do what you can do,' but I still felt a massive responsibility. I was the only cameraman down there. If I didn't get it, then that was all down to me. On that first day of new life, I filmed as much as I could. Personal discipline went out the window; I just filmed anything and everything I could see. Will probably wasn't pleased having to sieve through the mountain of footage but everything I saw down my view-finder was unique and I had to capture it, whether that be in chronological order or not.

That evening I spoke to Becky via a video-call to tell her what had been happening. She knew I'd been a little low over the winter and she could tell immediately that the presence of new life had lifted my spirits enormously. Entering a new phase of filming, I was excited to get back out there to see more.

Becky had been sending me videos from back in Northamptonshire. The summer that she'd been experiencing but I'd been missing was one of the warmest and driest she could remember. Having such a lovely, warm summer had made her single-parent life with young Walter so much easier. She could have him outside in the fresh air as much as possible and she was definitely noticing a difference from living in the North. The Lake District was notorious for its rain even through the warmer months, and despite it not affecting us when we lived there, now that Becky was being reminded of what a British summer *could* be like, she had started thinking about living elsewhere.

She knew this was a touchy issue with me. I'd grown up in the Lakes and loved it there. The freedom, the people and the scenery were unparalleled, and I was desperate to experience my youth again but this time with Walter. Without directly talking about the subject, Becky sent me advertisements for houses in the Midlands close to where her parents lived. I received a video she'd taken on her phone of a purple buddleia bush covered in Peacock, Small Tortoiseshell and Painted Lady butterflies. With Walter asleep behind her in his buggy, she joked about how good the weather was and how she thought she'd like to live there. I didn't take much notice initially, thinking it was a spur-of-the-moment thing, but when she actually began conversations with me regarding moving, I shut them down. I was too far from home, too detached from real life, to be making those kinds of huge decisions.

With the thought of an exciting and important few weeks with the penguins approaching, I put the subject to the back of my mind and forgot about it. I found it very difficult to talk about the future and make plans while down in Antarctica. Becky wanted to talk about what we would do when I got back and trips we might take but I found it impossible to think about anything but the situation I was currently in. I understood Becky had experienced a tough few months, but in different ways, so had I. With family around her spending so much time with our new baby, I was a little jealous at times of what she was experiencing and I worried about my return. Life was continuing without me and it felt strange to be so detached. I was grateful that days were getting busier, which helped to distract me from the worries of returning home.

The arrival of the newest members of the colony coincided with a week of some clear, calm weather enabling me to spend extended hours each day with the birds. It was an unbelievable relief to get a good spell of weather at this point and it allowed me to make the most of every minute with the birds. With more females due to return and take over the parental responsibilities from the fathers, there was a lot of behaviour to cover. During this period I quietly hoped for some little unexpected nuances that may add something special and stand out in the final film, but predicting these was impossible. The more time I spent with the colony, the higher my chances were of seeing something unusual happening.

Across the colony, more and more eggs began hatching. Almost every egg-laden male I saw had an egg with a crack in it, and witnessing those intimate moments when the helpless chick appeared became commonplace.

The next challenge was seeing a female returning to her partner and laying eyes on her own chick for the very first time. I needed a huge stroke of luck. I followed females patrolling through the main group calling and listening for a response; it was purely vocal recognition that would direct them to their chick and partner. With 5,000 birds to select from, who all sounded identical to me, I had no idea what differences they were listening for.

Males, meanwhile, continued to waddle around probably starving and desperate for their mate to return. The males looked so thin compared to the females. I wondered how long they could survive before their females returned. Would some of them have to make a choice between staying with their chick and starving to death or abandoning their chick and heading off to feed? Once their female did return, they still had a thirty-kilometre trek to get to the pack ice before being able to even attempt to catch some food. I don't know how they were still surviving.

Whether the males had a complete egg, an egg in the process

of hatching or a chick, they couldn't help but show off their prized possession to any passers-by. Standing side by side, males would lean back so far they'd rely on their stiff tails for balance. Stretching their bodies tall with their heads bowed towards their feet, they would roll up their feathers, revealing the contents of their brood pouch. They appeared proud and at times I found it easy to mistake a pair of males showing off to each other for the behaviour between a male and female pair. It was almost identical to when the birds were displaying and looking to pair up back in April; the males would lift their beak skywards, emphasising their golden collars. There was never any affection nor any aggression between the males and I found it heartening knowing this was how they operated. They were making life as easy as possible for each other, like they'd accepted the fact that facing the weather was a big enough challenge in itself without wasting energy fighting between themselves.

I tried my luck and spun the camera onto a female who was heading in my direction through the maze of males. I had no idea how long she'd been searching for her mate or when she'd arrived back, but I decided that following the mobile females seeking out the settled males would give me the best chance of capturing the moment. To my amazement, she stopped and called beside a male who stood right in front of me and immediately got a reaction. The male stood up and all of a sudden looked alert. Just like when showing off to his male mates, he leant back, vocalised and exposed his new chick, giving the female her first view. She became fixated, hysterical and couldn't seem to stand close enough. With her head bowed, she edged forward even closer to her partner and chick. Each step was exaggerated as her head bounced against her chest with each footfall.

With them standing next to each other, I could compare them both. The females had looked thin, weak and tired when they had left, but this female now looked incredibly fat and healthy. I knew the males had lost weight and deteriorated but until this moment, with a female to compare him against, I didn't realise the toll their ordeal had actually taken on them. It was a stark realisation of what the brutality of winter had done to the males. Having gone through the winter with them, it hadn't been obvious to me what damage they were doing to themselves. Their sacrifice of food for the purpose of egg incubation had resulted in each male losing almost half his bodyweight! This seemed huge, but it was their only option.

Leaning forward to properly inspect her chick, she seemed desperate to get hold of it. Just like back in May when the female had transferred the egg onto her partner's feet, it was now the male's job to transfer the chick onto her feet. His mammoth period of starvation was almost over. Advancing forward with her head low, she forced the male backwards. Having taken such tremendous care over the egg and chick for so long, he appeared understandably reluctant to let the chick go. It had been instinct since the day he was passed the egg not to let it out of his care and now that the egg had turned into a chick, the bond seemed to have become even stronger.

The chick's thin layer of grey down had now dried out and appeared to give it a little extra insulation against the cold, but it was still entirely reliant on its father for warmth and protection. After the female had made several attempts at knocking the male off his feet, he eventually gave in, stepping back and leaving the helpless chick on the ice. Shivering, the chick was too weak to even lift its own head. I maintained the camera's position, showing nothing in my viewfinder other than the wrinkly chick lying with its head resting on the snow. It seemed to be an eternity before the female's feet and beak entered the shot as she attempted to gather it up.

Just as the female shuffled forwards to collect her precious cargo, the exposed chick came under attack. Less fortunate birds close by, which had either returned and failed to find their partner or realised their chick or egg had died, had noticed the vulnerable chick. With strong maternal instincts kicking in, they were trying to snatch the helpless baby. A huge commotion ensued as other inquisitive birds joined in. Following the chick with my camera, I watched as it attempted to scramble to safety, using its feet to push itself along the snow. It was still tiny. As more penguins piled in, the chick was at serious risk of being crushed. It was heart-breaking but I didn't have time to think about that. I went into a zone, not really taking in what was happening in front of me, just making sure I captured the whole event in as much detail as possible. The chick disappeared under a mass of heavy bodies. Feathers fell to the ground as adults fought for the right to grab hold of it.

Within just twenty to thirty seconds an adult emerged from the scrum, shuffling towards me in a hurry to get away from the pandemonium behind it. On its feet was the chick. Covered in snow, with its mouth open, clearly exhausted, it had survived but it had no idea what had just happened. As the commotion subsided, I realised the bird carrying the chick was not its mother; it had been kidnapped.

As if nothing had happened, the scrum dispersed, leaving the two original, now chickless, parents wondering what had occurred. They kept inspecting each other's feet, wondering where their precious chick had gone. As with so many moments I had spent with the penguins, it was difficult to watch, and I felt so sad for the pair. After the enormous efforts both parents had made to get their chick to this stage, they had nothing to show for it. I was as confused as they were. Would their chick have as good a chance of survival being raised by a different parent? I seriously doubted it. Leaving the bemused couple trying to make sense of their empty brood pouches, the new owner of the chick shuffled off into the colony.

That evening over dinner, Will asked me how the day had been. 'Same old,' I replied, 'oh, but we got a kidnapping,' I added, shovelling pasta onto my plate. Will's face immediately lit up, part shock, part delight. It hadn't really dawned on me how incredible the behaviour that I'd witnessed that day had been. The strength of the maternal instinct in the emperors was widely known and recording occasions of adults fighting over chicks was on our wish list, but to capture a *successful* kidnapping so clearly with such a young chick was beyond our wildest dreams. As with most small events, I'd filmed it to the best of my ability, but it hadn't sunk in what it actually meant. These tiny moments, which had taken hours and hours of filming to capture, were the ones that would make our film stand out and that made all our time spent with the birds worthwhile.

Will called for a celebration. I got the impression he was slightly fed up of my low-key analysis of each filming day and my not-so-positive way of thinking, so he wanted to make the most of this moment and celebrate. Without making a thing of it, Will always ensured he praised me when I came back with something special and he made sure I knew how important these small moments were. He was well aware of the effort I was putting in down at the colony in incredibly difficult conditions. As ever, we discussed how we could improve on filming the behaviour if it happened again, to make such an event even more unique and special, but we didn't see it again.

It must have been hard for Will spending so much time inside waiting for Stefan and me to return each day. The majority of the time, his job involved sitting in front of a computer screen, but I was desperate for him to be experiencing everything in the same way I was.

With the weather improving we were making trips down to the sea ice almost every day. We started leaving our main skidoo sledge next to the ramp rather than towing it all the way back to the station, just parking it up for the night and reattaching it in the morning. So long as we had our emergency box with us, I wasn't worried. Despite the few days of calm weather, allowing me to capture chicks hatching and the behaviour associated with it, a couple of rough days did catch me out, preventing any filming down at the colony. I worried about the chicks and even though it wasn't the worst weather we'd experienced, its timing could have been better. After deciding it would be safe to leave a sledge next to the ramp, we returned the next day to find our sledge, containing our large custom-made wooden kit box, buried under a metre of snow. I should have realised that you could never predict what Antarctica would throw at you, and I should never have been complacent.

We tried to pull it out with our skidoos, but the sledge remained locked solid in the densely packed and frozen mound of snow. After half an hour of digging, we abandoned it and just grabbed the kit out that we needed for the day. We decided to finish the job of digging out the sledge on our way back after filming.

Returning to the colony after the weather system had passed over, I expected casualties. They were inevitable in such a harsh environment but it didn't make looking down at chicks that had succumbed any easier. Having not witnessed how these helpless individuals had lost their lives, I could only speculate, but with high numbers of females returning and chicks being passed between their parents, I guessed it was during the precarious handover process that most of the chicks had lost their lives.

As I did on most days, both to stretch my legs and to keep warm, I left my camera sledge in position and took a wide walk around the birds to observe them. Several tiny lifeless chicks poked out through the snow. I bent down to take a closer look at one. As if set in concrete, two rigid feet and a head facing skywards were all that was visible above the surface. The chick's eyes were half open and its beak was frozen in position, slightly ajar. With not a mark on it, the chick had simply frozen to death. Having spent a winter with the chicks' fathers, I knew what it felt like being cold, and it wasn't pleasant. I just hoped it had been a quick affair.

Within a few yards I stumbled across a much tougher sight: an egg with a chick half hatched. Lying on one side, the chick had managed to chip away the inside of the egg and had pushed the top of it away. Whether or not its father had then dropped it, I couldn't be sure. The chick looked as if it was sitting in an armchair, its tiny wings draped over the cracked edge; it had succumbed at the worst possible time. I tried to think how it had happened, visualising over and over the image I had recorded a month earlier of the males battling the elements. I couldn't imagine the panic it must have caused the father when the egg had fallen onto the ice. All his efforts, for nothing. It wasn't pretty, but it was natural, and it was my job to show what had happened.

Returning to my camera, I grabbed the length of rope attached to the two skis it sat on and pulled it over towards the dead chicks. Unfortunately, I had a few to choose from and having inspected them all, I went for the least graphic. Its tiny black eyes and beak were closed, its miniature feet were still on the ice and its head was in a natural orientation. That particular chick wasn't buried but was sitting on the ice in the position in which its tiny body had given up. I took some shots with my camera at the chick's level, showing the hustle and bustle of life continuing in the colony blurred in the background.

With my eyes now more tuned in, I started noticing more casualties closer to the adults, little inconsistent lumps that stood out on the surface of the ice. As I attempted to count them, a thin male emerged from the colony and started walking over in my direction. He had no chick or egg on his feet but walked with a stiffness as if he'd been restricted over the winter. As he strode closer, one of the little bodies lying on the ground caught his attention. Nosing it gently with the tip of his beak, he spun the rigid carcass around. After a few seconds he lifted his head and stepped away from it, but on several occasions spun around and returned. He couldn't leave it alone. Nudging it softly, he eventually positioned the frozen chick up onto his feet and covered the lifeless body within his brood pouch. With the body elongated in the position it had frozen, the chick's head poked out of the side above the adult's feet. He stood up and called as if the chick had just hatched. Watching him shuffle away into the huddle and begin showing it off to other penguins was one of the saddest things I'd ever seen.

Overcome with the emotion of paying so much attention to the chicks that had died within moments of hatching, I repositioned for the afternoon to film the enormous number of females that were returning from the sea. I needed to focus on something more positive. With huge bellies full of fish and squid, big groups travelling together started appearing through the light snow. Driving the skidoo away from the colony to watch females complete the last part of their journey to get to their chicks, I travelled north in the direction from which the females were coming. As the wind picked up, heavy drifts started reducing visibility and conditions became very poor. I began to wonder how the females were managing to get back to the exact spot they had left all those months ago.

With the storms the landscape had endured, the scenery had changed dramatically since they had left to feed back in May. Not all, but the majority of the shelf edge was now an extension of the sea ice with a ramp of snow having built up, merging the two into one. Icebergs that they'd had to travel past were still there but ice caves had been filled in and mounds of snow had built up their sides. The appearance of such structures would have been no visual help for the travelling females if that is what they were using to guide their return journey. It also baffled me that they hadn't been blown off course when feeding during rough conditions in areas where loose ice was moving in the sea's currents. A lot of birds were travelling during the day and under overcast conditions, so they didn't appear to be using the sun, the moon or the stars for help. It amazed me. There are still so many things that science doesn't understand. The colony had been in the same couple of square kilometres since forming in March, and despite having been pushed by storms and having rotated to avoid guano-stained snow, the birds had been extremely loyal to their patch of sea ice.

A group of returning females appeared through the drifting snow. With a patch of fog in addition to the snow, the visibility was no more than fifty metres. Conditions were tough, but they weren't storm force or unpleasant enough to demand an early end to our day. In the distance, the colony had disappeared into the mist. The females, paying no attention to me, my camera or skidoo parked close by, tobogganed straight past me at speed. I swung around to watch them travelling and realised they were going the wrong way. Or were they? The colony's position on that day was close to the edge of the ice shelf. I thought back to the time I'd filmed the birds laying their eggs. Then, the colony had been about half a kilometre away, and that was its last-known position, according to the females. They were heading straight for that portion of the ice. Was the birds' inbuilt GPS actually that accurate? Or was I just going mad? I found it unbelievable, so jumped on my skidoo and followed.

Arriving at the egg-laying site, the birds stopped, stood up and listened. With the sound of the colony now audible, they turned ninety degrees clockwise and headed straight for it. It seemed crazy that this was what they were doing, yet it made perfect sense. How were they managing to navigate? No matter how much was known to science about this aspect of their behaviour, it was astonishing witnessing it first-hand.

Following a successful day, which had included lots of mixed emotions, we finally packed up and readied ourselves to finish the job of digging out our sledge. Sitting on the skidoo, I sighed through my frost-encrusted balaclava. I felt exhausted both emotionally and physically. As I pushed the start button on the right handlebar with my thumb, nothing happened. I tried again; no response. For no apparent reason, another skidoo had broken down. It was the last thing I felt like dealing with, but needs must, so I grabbed a short length of rope out of the mountain rescue box, tied my skidoo to the back of Stefan's and prepared for the long journey back.

Deciding to free the snow-covered sledge the following day, we left it, spades still standing upright in the snow. I didn't have the energy to dig any more. Having to tow a skidoo *and* a sledge over an unploughed track would have been just too time-consuming.

Arriving back in the half-light, I dreaded telling the mechanic. He already had two skidoo engines opened up in the workshop being repaired. Telling him I had a third one for him, and that I wanted the ramp into the garage lifted so I could get another one out, didn't go down well. Feeling exhausted, I made myself a coffee from the machine in the lounge and went straight to my room.

My sleep pattern had only just started to return to normal, confirming that it was the disappearance of the sun that had caused the problem. Even though I was now getting some sleep, the lengthening days spent outside in temperatures that didn't appear to be increasing at all were incredibly demanding. I sat down in my chair at my desk with photos of Becky, Walter, Willow and Ivy covering the wall in front of me. Strewn across the tabletop was my fly-tying kit: feathers, strips of tinsel, thin copper wire and unravelled bobbins of thread. I'd taped an empty cardboard box that had once held six bottles of German beer under my vice for placing rubbish in, yet from the mess I'd made I clearly hadn't quite got the concept of why I'd put it there.

I put my hand in the pocket of my under-trousers that I'd been wearing all day and fished out a perfect emperor feather I'd picked up off the ice. Curved, white as the snow and with a grey and royal-blue tip, it was exquisite and I wondered what fly I could turn it into. Could I be the first person to catch a fish on a fly tied with an emperor's feather? I set to work. It was a new challenge that gave me time to myself. I'm sure the others thought I was being antisocial but being on my own was important to me and I needed that time to relax.

On 13 August, it was Becky's birthday, and I'd forgotten it! Having been so engaged with activity amongst the penguins, I'd not arranged anything other than a small card I'd left with her mother back in December. I felt so guilty and disappointed with myself for forgetting to get her anything from Walter. It was her first birthday as a mum, after all. Needing to burn off some of my anger with myself, I grabbed my unicycle and thought I'd give it half an hour to see if I'd improved at all. The few weeks I'd taken off from trying to ride it had done me good. It now seemed to come more naturally and I was managing a few metres at a time unaided. Each successful jaunt, however, tended to end with a fairly hefty crash. I was amazed my unicycle was still in one piece.

Just as I was getting the hang of it and beginning to get slightly carried away, the unicycle went uncontrollably from underneath me and I was left holding onto the scaffolding railing. At full speed, my machine piled into a stack of glass beer bottles that had recently been filled with some freshly brewed Antarctic beer. Using a kit brought down from Germany and Antarctic ice from outside, a couple of the guys had been brewing their own blend of beer. In one fell swoop, I'd smashed the lot! For all the wrong reasons, Becky's birthday wasn't one I was going to forget in a hurry.

As the weeks passed, I started focussing on more specific sequences that I was keen to obtain with the penguins. Capturing an aurora australis or southern lights display over an emperor colony had been one of my holy grails since I had started filming wildlife professionally aged eighteen. Being in Antarctica during the winter was my opportunity to witness this unique spectacle. From mid-spring through to mid-autumn, there is too much light in the sky to capture the aurora.

We'd been so far south for so long with so much darkness, especially through the winter months, that I was hopeful we'd be able to capture the penguins under the southern lights. However, although we'd had a few bright auroras over the station, by early spring we hadn't had a single opportunity to film one over the colony. The sun had now returned, and being close to the spring equinox, day-length was increasing rapidly. We were very quickly running out of time, and I was getting quite worried.

The auroras are incredible natural phenomena that are associated with the north and south poles and everybody is familiar with the image of an aurora's green hue across the sky, yet very few people get the chance to actually see one. We wanted to capture the penguins under the lights rather than just the aurora itself and for this to happen, many conditions needed to fall into place. First, for an aurora to happen, magnetic activity in the atmosphere needed to be strong. Second, we needed a clear sky with little cloud cover to be able to see the colours of the aurora, and third, we needed a decent-sized moon to illuminate the penguins sufficiently for our cameras to be able to see them.

On numerous occasions, two out of the three elements fell into place, but the three elements had never coincided. The requirement of a large moon limited us to just a few possible evenings each month. Day-length was increasing quickly and we were reaching a point when the light in the sky would outweigh the strength of the aurora; I was beginning to think we weren't going to get the shot.

On the evening of 4 September, an updated online forecast appeared. 'Solar storm is coming' read the title. The details of the magnetic activity around the sun and in our atmosphere were too complex for me to understand, but whatever had happened, an aurora was predicted within forty-eight hours. This was our chance. I immediately checked the upcoming weather, squinting my eyes as my computer screen refreshed the page, too frightened to look. It was forecast to be clear, and even better, we were one day off a full moon, which would rise at 7.30pm and set at 7am. Conditions were predicted to be perfect. All we needed was the aurora to peak during the night.

Preparing the cameras and making sure all our batteries were fully charged, Will, Stefan and I meticulously packed our sledges. We spent the afternoon resting at the station to try to make sure we were as fresh as possible to spend the whole night with the penguins. At dusk, we headed down to the colony. On arrival we unpacked a couple of cameras, sat on our skidoos and waited. The moon rose and the stars twinkled but there was no sign of any auroral activity in the sky. The anticipation was intense. We knew something would happen, we just didn't know when. Sitting and looking at the sky in the freezing air was incredibly peaceful. The temperature was around −25. As the moon began to fall back down towards the horizon, we were into the early hours of the morning and still nothing had materialised. With the first glimpse of daylight appearing in the sky, we knew we wouldn't see anything, so we headed back to the warmth of the station, disappointed that yet again all three of our key elements had failed to fall into place for us.

The next day was again crystal clear and, focussing all our efforts on filming at night, we rested in the station. Looking north from one of the windows, I could see the enormous icebergs in the distance. There was still a chance of succeeding; we just hoped that if an aurora did occur, it didn't happen during daylight hours when it wouldn't be visible. As we'd done the previous evening, we raced down to the penguins at dusk, getting all our gear ready before darkness fell. The sun set and was quickly replaced by a huge bright moon casting long shadows of the emperors across the ice towards us. With a colony full of adults brooding young chicks, nothing had changed. It was just as loud as it was during the day and with a luminous moon I could see the same icebergs in the distance as I'd seen from the station. In between, individual adults continued to make the thirty- to forty-kilometre journey to and from their fishing grounds in open water.

As darkness fell the silhouette of the ice cliffs in the southern portion of the bay behind the penguins became more pronounced. The sky was dark, but not completely black. The moon was rising to the east but now there was light appearing behind the cliffs that wasn't coming from the moon. Grabbing one of our cameras, which was more sensitive than my eyes, I took a photo and reviewed the image on the camera's screen. A band of light stretching across the sky from right to left appeared. It was faint but something was happening. Was this it? The sky was still clear and despite a slight breeze picking up, conditions remained ideal.

Within less than a quarter of an hour, the band of light had risen higher into the sky and had finally become visible with the naked eye. Although it was night, the large moon shining down onto the white snow and ice illuminated the landscape. I was wearing a head-torch but it was so bright I didn't need to turn it on. With the intensity of the light increasing, I placed a few cameras in position in case the aurora erupted. As well as several time-lapse cameras, which would record the event and replay it afterwards much faster than it actually happened, we had one incredibly sensitive camera that we hoped would film the colony and aurora in real time. This had never been done before and I knew if we could manage it, the film would hugely benefit.

Having set up a few time-lapse cameras around the periphery of the colony, I started filming a different group of birds. Already the sky was filling with several more colourful bands stretching across the horizon, appearing from east to west. It was obvious that, up in the atmosphere, the bands were circling the southern end of the Earth. There was clearly an aurora in progress but I still needed it to be stronger.

Before my trip to Antarctica, I'd only ever seen two auroras before, both from the top of my garden back home in the Lakes. The last one had been bang on midnight at New Year with Becky by my side. Other than that, the only experience I'd had of auroras was watching footage on television and a few previous auroras over the station, which had all been relatively weak. Each one had moved slowly across the sky but had never materialised into the type of bright, colourful display they are famous for.

However, this evening felt different. The activity in the sky was building and an explosion seemed imminent. Watching the image down the viewfinder of my camera, I could see the brightness increase, confirming that the aurora was already becoming one of the strongest we'd witnessed. In less than a minute, the long green lines blew up across the sky, pulsating and rippling. Showers of pink, emerald rivers and white flashes illuminated the sky. I'd never seen anything quite like it.

The speed at which the patterns changed shape and colour mesmerised me as the aurora flowed in every direction across the sky. Directly above my head, blue-green shards of light rained down. The atmosphere was full of life. Rushing around, I ran back to the skidoos to grab a different lens. Will appeared hypnotised, lying flat on the ice looking skywards. I had no time to ask if he was all right. Spinning the camera in every direction, I grabbed as many shots of the sky as I possibly could before focussing on the penguins.

Their demeanour hadn't changed at all despite the amazing light show happening above them. They'd clearly seen it all before. With a landscape full of explosive activity it seemed strange that there was no sound coming from it. The only sounds I could hear were emperors calling, the odd chick begging for food and the sound of light snow hitting my snow boots. If I'd closed my eyes it would have sounded like a normal day at the colony.

For almost half an hour the intense auroral activity continued above our heads before finally beginning to die down. Despite the decreasing intensity, green patches continued to spread out across the sky throughout the night. Realising we'd made the most of it, we packed up our kit and headed back to the station. Over a beer we spoke about what we'd witnessed and managed to capture. Having had my professional head on, concentrating on filming whatever I could, I don't think what we'd experienced had quite sunk in. It had been one of the most extraordinary and unique sights our natural world has to offer and we joked about how few people on the planet had ever seen what we just had. We could now add our names to a very short list of people who can say they'd been amongst an emperor colony illuminated by the aurora australis.

7.

A Spring Like No Other

It was mid-September and we were well into spring. Over the winter, there had been only two major storms lasting more than seven days that had affected the penguins. Now, surprisingly, just when I expected an improvement spring brought major disruption.

During the winter, I'd had countless days stuck inside the research station unable to get out to film, but the majority of those unsettled conditions were the norm for the penguins. There had been the one occasion when I'd returned to the colony to find a couple of dead adult birds and a number of discarded eggs, and the period over hatching when some new chicks had fallen foul of the atrocious conditions. But on the whole, mortality had been surprisingly light.

Now that the sun had returned, I expected the climate to have responded and improved. I desperately wanted to feel the warmth of the sun on my skin. But for weeks, the wind didn't relent. Despite endless periods of clear weather when the sun shone brightly down over the colony, temperatures only seemed to improve by a few degrees.

In addition, the blinding light created by the white landscape had returned and it was impossible to open my eyes without my dark glasses and tinted goggles. Although clear days *looked* pretty and more comfortable, in reality they were hell. Without a layer of cloud to insulate Atka Bay, cloudless days, just like frosty mornings back at home, were extremely harsh. It wasn't uncommon to experience fifteen to twenty degrees Celsius difference between a cloudy and a clear day and each morning that I saw the temperature display anything below −35, I dreaded going out to film.

Having overcome various kit malfunctions in the cold, there was one that recurred in any temperatures that hovered around the −40 mark. The fluid in my tripod head would freeze, meaning I couldn't move the camera at all, rendering the kit almost useless! Having a tripod head that locked solid was the first problem I'd encountered back in the freezer testing room in the UK. To get around this problem we'd had both our units sent back to their manufacturer to have the fluid replaced with something that wouldn't freeze at such low temperatures. When they were tested again they'd been given the all-clear but out in the field in *real* conditions, they were struggling.

On certain days I'd only been able to reposition the camera by unscrewing the levelling bowl underneath, pushing the camera in the desired direction and tightening it up again, annoyingly time-consuming and difficult in large mittens. I had no way of panning the camera to follow any action and although fortunately I had not missed any key shots, it was incredibly frustrating. Failure to record an important shot would, I think, have ended up with me launching the camera across the ice in irritation.

The long cold days were taking their toll on me and my patience was wearing thin. Unfortunately, it was a problem I just had to put up with as there wasn't anything I could do other than pray for slightly warmer conditions. Days with an actual air temperature of −40 had thankfully become less common, but the feeling that the wind chill gave me wasn't easing. Despite my expectations, spring didn't make things any easier for us. Each day, as wind speeds failed to drop below forty kilometres per hour, I experienced that stinging pain between my eyes I'd felt so often during the winter. I was always nervous of getting frostbite but so far I had been lucky and only suffered a bit of frost nip.

I wanted to be able to capture the new life within the colony in calm, serene conditions, but with such a prolonged period of unpredictable and windy weather, it was proving impossible. I'd had visions of the spring arriving and the birds, having survived the craziness of winter, now existing happily in an icy paradise with chicks playing and adults mingling, but that simply wasn't the reality. I had to show it as it was; I didn't have a choice other than capturing the moments as they happened, whatever the weather may be.

From the first hatching egg to the last, I'd spent almost a month filming the process and had been very lucky. A month sounds like a long time but it actually went very quickly. The birds had been cooperative and the time I'd spent with them had brought the reward of incredibly detailed images of unique events. With a constant stream of both males and females heading back and forth over the sea ice towards open water to feed, I'd quickly lost my ability to distinguish the difference between the two. It hadn't taken the males long to put on the weight they'd lost during their winter fast. Roughly every two weeks, the males and females swapped over, one taking charge of parental duties, the other heading off to collect food.

As the parents were returning with enormous bellies full of fish and squid with which to feed their young, the chicks were growing fast. They were still under the protection of their parents' brood pouches but were quickly running out of space in there. They had become incredibly fluffy, cute and inquisitive. Their large flappy wings draped over their parents' feet while their heads poked out from underneath the adults' white feathers and rested on the snow. Every now and then, they'd somehow pull themselves inside, rotate, lift their tail and excrete with extreme force onto the ice. Clearly, it was a priority not to mess up their parents' pristine plumage. Feathers in good condition were essential for the adults, who needed to stay perfectly insulated and warm both on land and in the water.

The chicks were real characters and even though they still relied on their parents, they were getting stronger each day. It was a privilege to spend so much time with them. A huge quantity of food was presented to the chicks with each feed. A lot of the time it was too much for the chicks, with half of it spilling onto the frozen ground. In no time at all the colony had become littered with lumps of frozen squid and in some cases whole specimens that barely had a mark on them. It didn't take long for each spilled feed to freeze solid and become inedible. It seemed like such a waste knowing what the adults had gone through to collect the food.

Keen to see exactly what they were feeding on, I collected what appeared to be a whole squid and took it back to the station to defrost. Sealing it in a clear sandwich bag, I hid it at the back of a shelf in a heated storeroom. I didn't think anyone would find it in there. I don't know why I hid it but I didn't want anyone interfering in case they didn't agree with me collecting it. It reminded me of when I was young: I used to find all sorts that I wanted to keep and my mum's freezer was the best hiding place. Luckily, she didn't often find out, and nor did anyone in the station this time. I wanted to defrost the squid to see exactly what it looked like. Its tentacles and head were slightly deformed. Not surprising, having spent a week inside a swimming then tobogganing penguin before being frozen.

The quantity of squandered items of food frozen across the ice amongst the colony made me realise how productive the ocean was and just what amazing hunters the penguins were. Despite not having managed to reach open water to see the emperors fishing, I noticed that they were all travelling in the same direction. With their social nature, many were in groups and I presumed were feeding together. Were all 5,000 penguins fishing at one time in a relatively small area along the coast-line? There is so little known about the penguins' behaviour when out at sea. I found it incredible just how much food the penguins were returning with. The ocean must have been heaving with life, a complete contrast to life above the surface, especially after an Antarctic winter.

Within the first month, the chicks had doubled in weight and were requiring almost a kilogram of food each day. They were still being looked after by a single parent while the other was out at sea fishing. The process seemed to be never-ending.

With the consistently strong winds, the landscape was proving a difficult place to work in. Like sand blowing on a beach, tiny particles of ice drifted across the ground. Despite being challenging and demanding, the ice particles constantly drifting through the air created unearthly and beautiful phenomena. When I looked towards the sun, it wasn't uncommon for it to be surrounded by a huge halo with intensely colourful 'sun dogs', or bright spots, around the 'compass points' of the halo.

Although withstanding such winds was what they had evolved to do, the chicks looked like they were having a tough time. Adults did their best to shelter them by turning their own backs to the wind and allowing the chicks to either climb into the brood pouch or tuck low in front of them. Birds in the centre of the group, who had the protection of the other penguins on the windward side, seemed to have it the most comfortable. Many of the chicks were now large and cumbersome, almost too large to fit in their parents' brood pouches, but there were still some extremely young, not-long-hatched individuals. Sadly, looking at these tiny chicks, I knew they wouldn't have enough time to get the feathers they needed to swim before the ice broke up and it made me realise just how fine the line is between success and failure.

With most chicks now larger and able to withstand the low temperatures, the parents became more mobile and less worried about dropping their young onto the ice. As the adults walked through the colony to stretch their legs, the chicks, still balanced on their parents' feet, always went with them. Although the chicks still restricted the adults' movement, it certainly seemed easier to get around than when they'd had an egg balanced on their feet. With each chick's foot balanced on the corresponding adult foot, the parent waddled slowly, stride by stride. Like children on stilts, it was comical watching the surprised chicks go along with their parents. The chicks' feet had the same black claws and grey scales as the adults', and seeing one on top of the other, the same except a fraction of the size, highlighted that in no time at all it would look just the same as its parent.

Within four weeks of hatching, the chicks had grown too large to fit inside their parents' brood pouches. They had grown strong enough and had produced so much fluffy grey down that being off their parents' feet and on the ice wasn't immediately life-threatening. Reluctant to leave the comfort their parents provided, they needed to be encouraged to move out. Parents opened their feet so wide that the chicks couldn't straddle the gap and were forced onto the ice. The parents stepped over them, often kicking and rolling them out into the real world. Slightly dazed, the chicks would immediately run at full speed towards their parents, forcing their head through their parents' feet and out the other side under the tail, leaving their body in position under the warm layer of feathers.

The chicks seemed shocked that they were being forced to move out but it wasn't long before they were making friends with fellow youths in the same position. Each day more and more chicks appeared to have been forced onto the freedom of the ice away from their parents and despite the initial shock they were soon enjoying life running around the colony through the maze of adults. Despite having a constant mist of ice blown in their faces, they looked happy out on the ice. They certainly coped with the non-stop onslaught of the wind a lot better than I did.

The more the season progressed, the more I expected a dramatic improvement in the outside conditions, so when the meteorologist announced one evening over dinner that he was expecting the strongest weather system of the year yet to pummel the station, it was my worst nightmare. Maybe I was being naïve, but having made it through a brutal Antarctic winter I really was expecting the spring to be much easier in terms of weather. The last big storm had presented such hostile conditions that I wasn't sure the adult birds could put up with anything stronger, let alone the newly hatched young. Just like their fathers, they'd started forming their own miniature huddles to keep each other warm but in wind speeds in excess of a hundred miles per hour, they wouldn't stand a chance. The following day, we finished filming before the weather was due to hit. I wished them well as I left. They seemed blissfully unaware of the storm they were about to experience.

As predicted, overnight the wind picked up and for two weeks it didn't ease. The environment was throwing one last gigantic hurdle at the birds before it allowed the tranquillity and ease of summer to take over. The station vibrated, our skidoos outside the front door became buried and our parked sledges blew across the ice. Everything we'd experienced came back in a more brutal way than I'd ever seen before. Temperatures weren't the lowest I'd seen but the wind was without doubt the strongest. Day after day, I longed for it to give in, knowing the damage it would be doing.

In the back of my mind I could picture what was happening. The unrelenting weeks of wind, although not storm force, had been slowly pushing the entire colony in a westerly direction towards the ice shelf. Although large ramps had formed up onto the ice shelf, large areas still hadn't been connected to the sea ice, and in between the sea ice and the ice shelf were death traps: enormous gullies with walls up to twenty metres high. The longer the storm lasted, the higher the chance of birds edging closer to these gullies and tumbling in. In the zero visibility that the winds were creating there was no way the penguins would be able to see the danger and the more I thought about it, the worse the images in my head became. I dreaded returning.

Throughout that couple of weeks, I kept myself as occupied as possible in the station, trying to take my mind off what was happening outside. Becky and I were able to speak several times every day and 'Pyramid Mountain' on the gym bike once again became an attractive and realistic option. It was certainly safer and less damaging than the unicycle.

Becky and her family had booked a sunny holiday abroad and were packing in preparation to fly. She listed items to me she'd had to pack now that she was responsible for Walter too, but revelled in the challenge of filling two suitcases: I dreaded their weight!

I'd never been super keen on going away on holiday when I wasn't filming. Working away all the time, all I usually wanted to do was spend periods at home sitting by the river with Becky and the dogs. But for the first time, a holiday in the warmth sounded quite appealing. It was Walter's first trip abroad and although he wouldn't remember it, it was yet another huge milestone that I'd had to sacrifice to be with the penguins. I'd received lots of emails from back home from overwhelmed friends who'd finally met Walter, all saying how perfect he was, and I was flattered at the number of people who had sat down and taken the time to send me an email. The community back home was a big part of my life. I knew almost everybody who lived in our village and all had personally wished me luck before leaving.

During my stay I'd been sending short monthly articles, just 600 words or so, back home to be published in the local newsletter. I felt it was important that I kept such an interested and loyal group of friends up to date, giving them the chance to experience my unbelievable adventure along with me, albeit from a distance. I missed them all just as much as I missed my family and was looking forward to sharing stories over cups of tea on my return.

My mum was on a holiday of her own over in France riding her bike up famous Tour de France climbs and I was receiving daily pictures from the tops of mountains such as Mont Ventoux and the Tourmalet. I started feeling a little trapped inside, not able to go anywhere while the storm blew over. I was witnessing incredible sights but also missing out on an enormous amount. I tied flies and, with the weather amazingly not seeming to affect our internet connection, asked everyone on the station if they minded if I streamed the Vuelta a España (Spain's equivalent of the Tour de France) live. Streaming video required almost all of the station's connection, making it difficult for others to use the internet during the couple of hours during which the Vuelta was on each day. But together with Dr Tim, who joined me on several occasions to watch the Vuelta, I was able to keep up to date.

On days when the others needed to be online to work, I

video-called Becky and got her to put the cycling on her television and place her phone in front of it for me. This didn't require as much of our limited internet capacity. The picture wasn't as good but at least I could hear the commentary. I sometimes watched for well over an hour and I wondered what she was doing in the background while I enjoyed the cycling. Being able to enjoy privileges such as the internet and connectivity seemed a bit strange at times. Living in such a hostile environment was widely known as being tough, but with such comfortable living quarters, morale could easily be restored after difficult days, keeping me keen to go out again into the painful outdoors to obtain film material.

After just under two weeks, which felt like a lifetime, a break in the storm opened up, giving us the chance to get back down to the birds and survey the damage. Will kitted himself up with Stefan and me, and we made our way over. I knew that what I was going to find would be ugly. The question was, just how ugly? In stark contrast to the previous weeks, it was calm and bright and for the first time since the summer I could feel heat from the sun against my skin.

Still a kilometre away from the edge of the ice shelf, the devastation started to become apparent. Like molehills in a field, hundreds of fluffy bodies lay strewn across the frozen ice sheet. Having been blown huge distances away from the colony, the helpless penguin chicks, who had so recently lost the protection of their parents' brood pouches, simply didn't stand a chance.

Rather than stopping the skidoo, I slowed and weaved my way through them, taking extra-wide turns so as not to clip any bodies with my long sledge. There was a tremendous amount of noise coming from the colony, which now spanned a huge area in fragmented groups. Some birds had been pushed so far by the wind they'd been forced up some of the naturally formed ramps and onto the elevated ice shelf. With so many birds, we parked our skidoo and walked over to the edge. What we saw was nothing short of utter carnage.

The gullies in between the ice shelf and the sea ice had doubled in size and depth and were full of emperors and their chicks. Some adults' heads hung low, weighed down by enormous blocks of ice that had matted in their feathers. In a small depression in the ice, a huddle of forty or so chicks lay dead. Their grey coats stained with brown and green penguin excrement, they appeared to have been trampled by the unstoppable rotating colony during the peak of the storm. Some chicks lay on their backs, still alive but stuck to the ice. Waving their feet in the air, they were attempting the impossible task of righting themselves. Parallel to the ice shelf, along the bank of snow that had built up, a huge narrow crack where the sea ice had buckled under the pressure had opened up. Looking down into its dark depths, I could see chicks desperately trying to scramble out, attempting to grip onto the near-vertical walls of ice.

Directly below me, about five metres down in a large crack, an adult lifted its head and looked me in the eye. On its feet was its chick. I couldn't bear it. I had spent so long with the penguins and gone through almost every moment of their breeding season with them, and they felt like family. Seeing them in this state and not being able to help was heart-breaking. I didn't know where to look or what to do. The incessant conflict between the weather and the penguins had reached such heights it seemed to have boiled over and got out of control. In tears, I walked over to Will and Stefan, who were equally upset.

No matter what our feelings were, I had to remain professional and record what I was seeing. With a rope and a harness, the guys lowered me and my camera down into the end of the gully opposite to where the birds were positioned. Rather than abseiling, I sat down and slid on my backside. Despite having been formed initially by snow build-up, the walls were now rock solid and even using the heel of my heavy snow boots, I could barely make an indentation when I kicked into them.

The penguins, trapped at the bottom, seemed to have already accepted defeat. Shaded by the near vertical walls of ice that rose around them, they huddled together with their heads down to keep warm. I set up the camera at the same level and used my long lens to get a better look at them. The floor of the gully was flat and seemed to be formed from the sea ice, without any snow layered on top of it. I had to be careful where I was standing, as to move around freely I'd disconnected myself from the rope that hung down. Perished chicks, each covered in a crust of penguin excrement and ice, littered the ground around the huddling adults' feet.

One bird, standing separate from the small huddle, was covered in blood. The top of his head was so matted with blocks of heavy ice that he had attempted to shake them off, in the process tearing away feathers and skin from his scalp. It was so raw I could almost see his skull. Water freezing and becoming ice is normally a straightforward occurrence, but in a place where the ice simply doesn't melt, the consequences can be catastrophic.

As overwhelming as it all was, I attempted to block out my feelings and document as much as possible. Having managed to cope with death in the colony, I was slightly startled by my reaction to the drama that was playing out in front of me. I think I was struggling to come to terms with witnessing the suffering that led to death. Looking down at dead, lifeless bodies had become a frequent occurrence, but standing and watching them perish in front of me, unable to do anything to help, was incredibly distressing. With eyelashes weighed down by heavy crystals that had formed from my own tears, my eyelids were sticking together.

So many chicks having succumbed in the colony, hundreds of adults now roamed free, vocalising, searching for their missing infants. The gully had attracted a lot of attention from calling individuals and some adults peered over the edge, gauging the best way in. To my amazement, penguins started scaling down the seemingly impossibly steep walls with ease. With no chick or egg balanced on their feet to worry about, it seemed effortless. Using their beak as an ice axe, their wings for balance and their feet and sharp claws as extra grip and power, they traversed the ice at incredible speed. Having descended from the top, some adults who'd been calling seemed to either be looking for their partner or for their lost juvenile. It was a painful sight; desperate to find their family, they were inspecting every last corner of the colony.

Some adults who were huddled together at the bottom of the gully still had chicks tucked away in their brood pouches, unaware of the danger they were in. These chicks must have survived the enormous tumble into the gully with their parent.

Whereas before, the chicks had represented new life and hope in the colony, they had now become a deadly burden on their parents. The adults were simply unable to climb out of the gully while caring for their helpless chicks. It was a case of both dying a long and painful death together, or the adults saving themselves by sacrificing their chicks. It was a harrowing predicament.

For hours I stood, practically motionless, observing what was happening. I wanted to understand the intentions of the penguins so I could film their true behaviour. As with previous incidents like the snowball-egg, the kidnapping and their relocation on the sea ice, this was one of those moments that stood out, and if I filmed it respectfully and truthfully it would be a powerful sequence in the final film. It would show the true brutality that emperor penguins have to endure year in, year out, of which most people, including myself, had been unaware.

Halfway through the day a couple of penguins that hadn't moved at all since my arrival lifted their heads and started paying attention to their surroundings. Their chicks had periodically poked their heads out from underneath their parents' feathers and begged for food, but I'm not sure whether they were able to provide for them. With very little warning or emotion, these penguins looked as if they'd had enough. Stepping away and abandoning their chicks, they made the ultimate sacrifice. There was no looking back; they did not appear to have any second thoughts. They knew what they were doing. For the helpless chicks, their first few premature steps onto the ice would be their last.

It was clear, however, that for the remaining birds, abandoning their young was absolutely the last resort, and for the time being their strong parenting instinct made them unwilling to make that choice. I checked the gully from every angle, exploring possibilities for the penguins and their chicks to escape, but there did not appear to be any; it had become an icy prison. With the mountainous barrier of ice that hung over them, there was only one outcome.

But one mother refused either to sacrifice her chick or accept her fate. Shuffling along the base of the gully, she surveyed her options for freedom, every now and then bending down to check on her young chick. It was obvious what she was thinking. She wasn't giving up. Inch by inch she began her ascent up the initially shallow walls. Her claws crunched into the ice and sank a couple of millimetres into the snow. She started zigzagging to reduce the steepness, but to no avail, losing her grip and sliding back to where she'd started. She went again, taking a slightly different route, using untouched ice to aid her grip. I'd never read anything in my research about emperor penguins climbing. I couldn't believe what I was seeing. They were built for the ocean and swimming at speed to hunt and avoid marine predators, and even though they usually appeared clumsy on land, this female appeared in total control. Utilising her beak to full effect, she began to stab it with full force into the ice to give her ankles and toes a rest. The chick had swung around and was nestled in her brood pouch head first. Whether this made a difference or not I was unsure, but she was managing to make progress.

For over an hour the progress was painstakingly slow as she edged her way closer to the top. I longed for the remaining birds at the bottom to copy her; normally, they followed suit with all their behaviour in and around the colony, and I was so surprised they didn't this time. The huddle of birds in the gully didn't even raise their heads. It was heart-breaking to watch them; they looked as if they had given up all hope of survival and had resigned themselves to a slow, painful death. The individual attempting the seemingly impossible was a one-off and as she neared the final lip at the top, I reconnected my rope and quickly pulled myself out to get a different angle on her potential escape. Although I got out of the gully in the hope that she would reach the top, I honestly didn't believe it was going to be possible. No other bird had even attempted to get out with a chick on its feet, let alone nearly reached the top.

I got out of the gully as quickly as I could, aimed my camera at the lip where I hoped she would appear, and held my breath. Would she make it or was she sliding back down the almost-vertical wall as I sat there and waited? I couldn't bear the thought of her failing now. I doubted she would have the energy to try again; this was her one chance. Ten minutes went by and still she hadn't appeared. I dreaded looking down to see her back at the bottom having to try to go through the whole process again.

Then, all of a sudden, the top of her head poked up from beneath the sharp edge. Stretching her neck high she could finally see the colony in the distance and had something to aim for. I buried my eye in the camera's viewfinder and watched her final effort, desperate for her to succeed. She looked physically exhausted but unbelievably determined. Ramming her beak into the flat ice over the ridge, she gave it one last push. Her wing tips pressed against the snow as she rocked from side to side with each foot creeping up the final few centimetres. The sharp ridge at the top of the gully was her last obstacle and without withdrawing her beak, which she was using as an anchor, she used every muscle in her neck to pull herself over the edge and out of the gully. Weak but safe, she was out, and so was her precious chick. I watched her shuffle back into the main group, the memory of what had just occurred apparently erased from her mind. With a wave of relief it was time to pack up and get back to the station. The weather was once again starting to move in; she had made it just in time. I could only hope that the winds wouldn't be as strong this time and that more birds wouldn't get blown into the gully.

That evening and the following day, we were unable to get back to the colony. So many things went through my mind. While down at the colony I'd had the distraction of filming, which had helped lessen the effect of the graphic images in front of me, but back at the station there was no such distraction and I couldn't help but think about the penguins constantly. I thought there must be something I could do. As a team we spoke in depth about what we'd seen and whether it would be right to try to help. To me it was a no-brainer; the reality was, I didn't think I could live with myself if I didn't do anything, but I had to be sure everyone felt the same. We all decided that, depending on the situation when we managed to return, we had to step in and intervene in some way to try to help.

The next forty-eight hours were tough. We were stuck in the station and all I could think about were the birds that were trapped in the gully and what they were going through. I constantly checked the weather and wind speed outside; it was up to ninety kilometres per hour again and I knew the birds would be struggling. After forty-eight hours, thankfully, the weather broke and we could get back down to the colony.

The situation hadn't improved; if anything, the gully had engulfed more victims. I couldn't help but ask why they were putting themselves through this. Breeding in one of the harshest places on Earth where no other life could survive had its advantages of course, in that the emperors could bring up the next generation without the threat of predators; but in a way they were having to fight a much bigger, more ferocious and terrifying predator: the Antarctic weather.

Peering down onto the group of birds and frozen chicks at their feet, it was clear that, since we'd left them, more birds had made the decision to abandon and sacrifice the life of their young to save themselves. We couldn't let that continue to happen. From the bottom, at the same end of the gully I'd filmed from, we set to work with big red shovels. The only thing we could try was to dig into the slope and work our way along, inclining it gently upwards. Slowly, we formed a narrow but much shallower ramp that they could potentially shuffle up with their chicks.

Large, heavy blocks of snow slid down the walls of ice as we smashed our way into the slope. Feeling a sense of urgency, it wasn't long before we were all sweating. A thin crust of ice formed across my stubble as my sweat froze. Soon, with just a couple of spades, we had made a ramp. I tested it myself, shuffling my own two feet up it to ensure the slope wasn't too steep. The ramp was just wide enough for my large orange snow boots and as they slid along I compacted and kicked away the loose snow from its surface. I wanted to make sure it was as smooth and obstacle-free as possible so the penguins would have an easy route to take.

I glanced across and wondered what they were thinking. It was so easy for me to imagine them seeing this escape route and taking advantage of it, but I knew it wasn't that simple. Were they clever enough to figure it out? Did they know what we were trying to do? Looking back down the ramp I was proud of our work and felt slightly easier knowing we'd done something. Whether it would work was another thing and I wondered, if they did show some interest, how long it would take for them to figure it out.

We cleared the ramp, tied our shovels back onto the skidoo and repositioned ourselves on the ice shelf further along, away from the gully, to watch from a distance. We'd done our bit; it was now up to the birds. Before we could even sit down, a couple of individuals headed to the end of the gully we'd been working in. Their inquisitive nature took over and they just couldn't hold themselves back from investigating what we'd been up to. Once at the base of the ramp, the first bird simply didn't stop. Slowly shuffling one foot after the other, he carefully scaled the incline all the way to the top. Having walked himself to safety, he attracted attention from chickless individuals at the top, who were, unbelievably, just as keen to get down into the gully as the trapped birds were to get out.

One by one, each bird became aware of the possibilities the ramp offered and by following the birds in front, they began to edge themselves to freedom. As I watched, I felt proud and emotional. I also asked myself, did this happen every year? Had we not been there, how many birds would have perished? There had been so many penguins and chicks that I'd been unable to help; there was no way I could have walked away knowing I hadn't done anything to help the ones I could.

I understood that our actions would be seen as controversial, that some people would claim this was part of the 'natural process', but being there I also saw the other side of it. The emperors and their chicks hadn't been pushed or chased into the gully by a predator, and their slow, lengthy death wouldn't have immediately benefitted any other life form. Apart from the emperors, there was no other life there. They would have died a long and painful death from starvation and hypothermia, eventually being buried under heavy snow, and we couldn't just sit and watch that happen. The help we had given them was indirect. By digging a shallow ramp, we'd given the penguins an option, a way out, but we had left the decision to them. They didn't have to use it, but they did; they saved themselves.

It was an incredible feeling of joy and relief watching them line up along the ramp and return to the colony. Having had such low moments through the coldest and toughest few months of our stay, it felt like a massive turning point.

Becky and Walter had been away on holiday for only a few days by that time, and images and videos of Walter asleep shaded on his sun bed dominated the message feed on my phone. As I was in −30, they were experiencing +30 degrees. Walter appeared to be loving the warm weather and children's swimming pool with his Uncle Rob, Becky's brother, and even though I was grateful for all the help they were giving Becky, I couldn't help but feel a little jealous. I worried that, despite Becky's best efforts, Walter would think his uncle was actually his father as he saw so much of him, but I didn't dare say anything.

Despite having to deal with difficult feelings, watching videos of Walter showed me that he was a real character. The more photos and videos that Becky sent me, the more I felt like I knew him. He seemed to be enjoying the stories I'd recorded, falling asleep while they played in the background. Apparently, that meant they were perfect.

One day, when I'd felt a bit down, Becky tried to cheer me up by reminding me that my birthday was just a few days away. It hadn't crossed my mind; not wanting to make a fuss, I'd not made anyone at the station aware that it was coming up. A few evenings before 17 October, I decided to bake a cake for everyone as a dessert after dinner. I had been really missing sweet foods as dinner rarely included a dessert; the best I got was a small yogurt after a meal.

Being an old-fashioned Brit and with the added bonus that it was easy and quick to make, I mixed together ingredients for a Victoria sponge. I loved baking back home. As was usual in the Antarctic, I'd had to make an alteration to the recipe, as having been isolated for almost eight months by now, we'd unsurprisingly run out of fresh eggs. After a bit of research I discovered that apple sauce apparently made a reliable substitute for eggs so went with it and to my surprise it worked. I demanded the team followed my tradition by accompanying their slice of cake with a cup of Earl Grey, and we all sat together and talked through my methods. It was lovely to have moments of togetherness on the station.

A few days later I woke up a whole year older than I'd been when I'd gone to bed. As usual I breakfasted on my own, kitted up and headed down to the colony with Stefan. I liked to have breakfast alone. It was nice to have a bit of peace and quiet in the mornings to get ready for the day. We had started to run out of things by this point and I generally had the same thing for breakfast every day: a defrosted bread roll, toasted and buttered, with a little cherry jam. To be honest, there wasn't much else available!

I found the food situation in the Antarctic tougher than I had expected. I'm generally not fussy and will eat anything but I missed fresh vegetables and salad so much. Lunch was the only hot meal of the day down there and as we were generally out working, we would miss this meal. I lived in hope that they would save us a proper meal each day, but it rarely happened. As I didn't want to rock the boat, I just kept quiet. Dinner was regularly cold as the chef was reluctant to prepare two hot meals a day and it didn't take me long to get fed up of defrosted cold meats, pickles and cheeses. What I would have done for a piece of broccoli!

Back down at the colony, the recent carnage from the gully was long gone from our memories. Although the gully was still there, thankfully it was empty. To grab some close-up shots of the perished chicks at its base with no prospect of disturbing any adults, I headed back down into the gully with my camera. Being so deep down, the sound of the colony on the ice above was all but blocked out and it was remarkably peaceful, despite the horrendous scenes I'd witnessed there just days before.

Back up above the gully, with the colony behind me, I located the huddle of dead chicks that I'd presumed had been trampled to death. They were yet to be buried under snow so I took the opportunity to document the appalling sight. As I set up to film them, I could hear the sound of chicks begging for food. It was muffled, but I knew immediately what it was. I looked around me but there was nothing. I was confused. Independent chicks that had been left alone by their parents hadn't stood a chance in such awful weather. The pile of dead chicks at my feet demonstrated that.

Assuming my ears were playing tricks on me, I crouched down to begin filming, but the muffled sound only became louder. I attempted to pinpoint it. With no clues whatsoever on the surface of the snow, I came to a conclusion that the sounds were coming from beneath. I took my radio out of my pocket, laid it on the ground to mark the position and returned to the skidoo to grab a shovel. Using it as if it were a trowel, I carefully started to scrape my way into the snow. I stopped for a minute and listened, but all had gone quiet. After a couple of inches, I broke through. The hole I'd created opened up into a small chamber and inside were four emperor penguin chicks, unharmed, sitting in a huddle together. During the storm, snow must have built around their mini huddle and buried them alive. Fortunately, this small group had avoided being trampled. As they had stayed close together, their body heat appeared to have formed smooth but solid walls around them. They were moving freely but space was so tight that they only just fitted.

Being careful not to touch them with the shovel, I scraped away and enlarged the hole I'd already made. They seemed surprised but keen to get out. I couldn't believe my eyes. The four young chicks had been on their way to certain death, not from the cold but from starvation. But on the other hand, being buried under the snow had undoubtedly saved them from being trampled to death. Watching them climb out, shake themselves off and make their way back over to the colony was one of the best birthday presents I could have wished for.

Back at Neumayer, Becky had been on the phone to Will to tell him it was my birthday, but neither he nor Stefan had forgotten. Over a slice of cake and a cup of tea, they presented me with a birthday gift. Having had to get inventive, they'd used items of broken kit we'd accumulated to create a collage on a sheet of wood we'd once used as a camera box. Screws, snapped cables, part of an un-repairable viewfinder, a bracket from a sledge I'd snapped and recycled electrical tape had been used to make a work of art: a three-dimensional image of an emperor standing proud on the ice under the sun. It was something to treasure forever!

Will had also made me a Battenberg cake. He had planned to make me a Victoria sponge but, to his annoyance, he'd had to have an emergency rethink after the one I'd made a few days before. Becky had told him that Battenberg was my favourite. I was incredibly impressed.

With the emperors' story finally coming together, there was another crucial event I was keen to witness: independent chicks braving a storm. Unlike when I'd filmed the males huddling against the storm-force winds, filming chicks in bad weather didn't require as much planning. Although the weather would need to be bad, it wouldn't need to be anything like as severe as it had been when we filmed the males, so the risk for us as a film crew was lower. All I needed was some poor visibility and a bit of drifting snow, comfortable enough for my acclimatised body to film in, but still potentially life-threatening for the small penguins.

For weeks, they'd been ganging up and tearing after each other through the colony and their confidence had definitely grown. Being independent and not so reliant on their parents had allowed both adults to head off to sea to fish at the same time, so it was more important than ever that the chicks stuck together.

I'd become very attached to them, especially after recent events; they were just as, if not more, inquisitive and friendly as the adult birds. Amazingly, they could identify their parents by their call alone, but they were so independent now it became rarer and rarer to see a chick with its parents. Even if the adults were back from the sea, the chicks didn't take much persuading from their mates to disappear through the mass of penguins and leave their parents behind, once they'd had a good feed, of course.

Despite at this stage relying entirely on their parents for food, the young penguins still had a tendency to follow other adults and it was common to see them attempting to squeeze under strangers' bellies for warmth, even if they didn't fit any more. Adults by this stage had lost that instinct to want to care for the chicks and didn't seem to tolerate little ones trying their luck. They were never hugely aggressive but meaningful prods with their beaks and backing away from the chicks ensured the message was conveyed. It was comical at times watching juvenile emperors test several adults one after the other, but when all the chick wanted was a bit of love and protection, I felt sorry for them.

With some heavy drift and a bit of fog reducing visibility, conditions started to become challenging for the small chicks. For me, the overcast conditions lifted the temperatures to around −20, making it feel tropical compared to what I'd been used to. Because of this, the emperors were dispersed across an enormous area of sea ice, and with the fog I couldn't see them all at once.

Filming the northernmost group of a couple of hundred birds, I began to see adults fresh in from the sea with bellies full of food arriving through the mist and starting to call for their chicks. In amongst the group, approximately thirty youngsters huddled in the centre, protected from the wind by the larger bodies of adults. As more adults arrived, equal numbers were leaving, starting the thirty-kilometre trip to the edge of the ice.

As a large bird passed through, my eye was caught by a small chick following it. Whether it was its parent or not I had no idea, but it seemed intent on staying with the adult. The adult either didn't know the chick was following, or didn't seem to care, and dropped onto its belly to toboggan the initial part of the journey. Speeding away, it left the chick in its wake. In that kind of misty weather, following an adult was a risky move, but the chick's powerful instinct to stay with the adult meant it continued to follow. The adult disappeared, none the wiser, but for the confused chick, it was a disaster.

As I looked back at the group of adults, it dawned on me how different my perspective was to the chick's. For me, at six foot tall, the drifting snow didn't present a problem. I struggled to see my feet at times as ice particles blew over, but for a chick no taller than thirty centimetres, it was a different world. I bent down to see its viewpoint to get a feel for what it was experiencing and immediately I felt disoriented.

The chick I had been watching had almost disappeared into the mist. I could hardly see for ice crystals hitting my eyes and my hat and balaclava were leaking snow. Watching the chick travel away from the adults across the sea ice, all I wanted to do was round it up and point it in the right direction. To me it seemed obvious; I could see the adults one way and I could see the chick walking in the other direction, but there was nothing

I could do. Unlike the certain death that the gully had presented, I couldn't be sure of the outcome here.

I picked up my camera and followed the chick as it continued to travel north, away from the protection of the colony. Eventually, I lost sight of the birds behind me; I also lost sight of my skidoo and Stefan. Out of the blizzard, a returning adult spotted the chick and rose to its feet. To my relief, the lost young penguin started to follow it. Whether the chick knew which way the adult was going I wasn't sure, but so long as it didn't lose sight of the adult this time, it would be led back to safety. The adult emperor remained on its feet rather than lowering itself onto its belly and as it strode towards the colony it repeatedly glanced behind. I couldn't help but feel it knew the chick was in danger and needed help. Every so often it paused, allowing the chick's short footsteps to catch up. There had been no reaction from the adult in response to the chick when it called, so I was positive it wasn't its parent; it was just a more tolerant individual who knew that it wasn't the time to be rejecting a curious youngster.

As they both returned to the safety of the colony, the adult didn't stop. It appeared to know its job was done and that the chick was safe. As the chick shook itself off and rejoined the huddle, the panicked look seemed to drain away from its little face. In what was probably just another everyday circumstance, I again felt quite emotional. I don't know what this little bird felt like, but it appeared to me to be another demonstration of the penguins looking after one another to survive. At least that's what I liked to think.

I thought back to when I was in a similar position trying to return to the station after filming in the storm. I'd had no one to follow. All I'd had was a tiny piece of technology and two colleagues, and I had been terrified. On a regular basis, from early in their lives, the penguins were up against it and every choice they made had consequences that could mean the difference between life or death. The chicks were less than twelve weeks old and those that were still alive had overcome unbelievable hurdles. I looked at the parents and began to truly appreciate the difficulties they'd overcome to make it to adulthood, let alone to parenthood. The odds really were stacked against them.

As the weather improved and became more consistent I started thinking about something I was hoping to do in Antarctica: getting to the ice edge where the penguins were emerging out of the crystal-clear blue waters. According to satellite images, the sea had already started to break up in the north as temperatures slowly rose and the nearest patch of mobile pack ice where the birds could access the water for food was about twenty-five kilometres away. Having seen the sea ice in the bay break up so much more quickly than expected at the beginning of my trip, I hadn't managed to get to the edge as planned. The winter conditions had been too unpredictable for it to be safe to travel to such remote areas, especially across a frozen ocean whose depth of sea ice I didn't know. Looking north every day towards untouched icebergs, I longed to travel across the unexplored landscape. Only the penguins who had travelled across it knew what it looked and felt like, and I sensed an urgency to go myself before the summer temperatures started to break the ice up again.

Six enormous icebergs sat as sentinels in the distance, having been locked in the ice since the sea had frozen. The penguins had spent two months making the journey, sometimes taking different routes but always heading in a northerly direction away from the colony towards the water. On a calm day, Will, Stefan and I left the colony behind on our skidoos and started following footprints. The sea ice, unlike our ploughed, marked track, was incredibly rough and uneven and making our way across it was time-consuming.

I'd gained a reputation for travelling slightly more quickly than perhaps I should have on my skidoo. Although I am not normally a speed demon, in Antarctica I found myself going more quickly than I normally would so that I could maximise my filming time in each location. Despite priding myself on being the lead vehicle and picking my route across the smoothest areas, on some occasions, especially in difficult conditions, I guided myself into difficulties.

I led the way, with Stefan behind me and Will behind him, cruising across the snow. In some areas it was incredibly fast and I felt as if I was floating, but in other areas, where icebergs had affected the flow of wind, it was covered in ridges and solid icy lumps. I felt as if I was back home on a rocky fell side having to negotiate my way through the jagged boulders. The further north we travelled, the more uneven it became. Gliding across a patch of undisturbed snow, my concentration drifted.

Suddenly, I slammed into a solid lump of ice. The front of my skidoo left the surface and rose into the air. I just managed to keep hold of the handlebars, but my head flew forwards, smashing my goggles into the skidoo's stem. Luckily, the skidoo landed upright, but the sledge was on its side and carved itself into the snow as I brought the machine to a standstill. Both guys stopped alongside me, having diverted around the lump of ice, and hopped off to right my sledge. I jumped away in shock, shook myself off and looked back at my machine, grateful I wasn't injured. Just like when I'd damaged skidoo number 10 at the North-Eastern Pier, the rear of my machine appeared lower and on closer inspection a steel shock absorber within the suspension spring had snapped clean in two. I'd knackered another skidoo!

Will was understandably frustrated. 'Lindz, you've gotta slow down!' he exclaimed. It hadn't been speed that had caused it, just lack of concentration, but five kilometres out onto the frozen ocean, my skidoo was stranded. Despite it being an accident, I felt guilty. More work for the mechanic and with the summer season approaching, it was just when he didn't need it. Being out on the sea ice, there was no way of recovering the machine with a snowcat; it was down to us to get it back to the safety of the ice shelf where it could be picked up.

For Will, this was his first day out of the station in a couple of weeks, and he was as keen as me to travel north. But being the true professional that he is, Will knew that if we swapped skidoos, Stefan and I could continue our day filming while he attempted to get the broken skidoo back to the station. The weather was settled with a favourable forecast so the three of us didn't have an issue splitting up. But would the skidoo move? I lay down and, while Will lifted the rear of the skidoo, I placed the snapped ends of the axle back together like two pieces of Lego. Will slowly lowered the rear of the skidoo before he let go. The axle held. So long as he didn't travel faster than five kilometres per hour, Will would be able to drive it back, saving me the embarrassment of having to radio for help. I unstrapped my camera box, unscrewed its bracket and moved it across to

Will's skidoo. 'Good luck,' he said as he began creeping back along our skidoo tracks towards the station. In the distance, the station looked within easy reach, but realistically it was an agonising four-hour journey.

The mishap had taken half an hour out of the morning but Stefan and I continued north following tracks from birds that had headed for open water. Along the way, penguin footprints stood proud of the snow like strange mushrooms, where the feet of marching emperors had compressed the snow and then the softer snow surrounding the footprints had been blown away and eroded. As I spun around and surveyed the landscape, I couldn't see a single bird. With the speck of the colony way behind me distorted in the rippling haze, the never-ending blue-and-white vista appeared more like another planet than ever before.

As the boulders of ice became more impassable I began to realise that getting to the ice edge was becoming an unrealistic prospect. I'd grown used to witnessing incredible events and pieces of behaviour in the penguins' life cycle and making it to open water to see them feeding and leaping out onto the ice was the last piece in the emperors' life jigsaw I needed to complete the picture. Through my ten or so months in Antarctica, I'd seen everything else an emperor penguin did first-hand, and more. My personal goal of being next to swimming penguins at the edge was ambitious, and after everything I'd seen, I was being greedy wanting more. I came to the difficult decision not to go any further. The skidoos and sledges were being pushed to their limits and I didn't dare risk breaking another, especially twenty kilometres away from the station.

Turning off the engine, I accepted defeat and pulled out a flask of hot chocolate I had in my backpack. It was the first time since before the winter that I'd used my flask, the liquid staying warm for barely a minute once in my cup. We grabbed some shots of an isolated group of penguins travelling round the icebergs as they transitioned to open water. It was nice to be filming, a welcome break from driving. I walked over to the nearest iceberg, a dark-blue jagged mass whose wall resembled a scrunched-up piece of blue paper. The sea ice on the sheltered side, which had been protected from the wind, was lumpy and rugged and one-metre-high blocks of ice lay strewn across the surface.

I pulled out Walter's penguin from my pocket; it had come everywhere with me since he was born. To keep it clean and dry, I'd put it in a spare dog poo bag that had travelled to Antarctica hidden in a trouser pocket, left there from walking Willow and Ivy back home. Sitting it on a block of ice with the sun's halo surrounding it from behind, I snapped a few photos for Walter's album to prove his toy penguin had travelled to the bottom of the Earth. It wasn't long now before we'd both get the chance to meet him.

As time was running out, there were a few things I wanted to do before leaving the bottom of the Earth. I was keen to return with images of items I'd travelled down with, to prove where they had been: my local cycling club's jersey and a 'bag for life' from the local village butcher. He loved seeing images from far-flung locations around the world that his bags had reached.

Back home, an annual bike ride was taking place in the village in memory of a friend who had died a few years back. I'd always made an effort to take part in the bike ride and although I couldn't make it this time, I wanted everyone taking part to know I was thinking about them. So I made it my mission to take the club jersey down to the colony and grab a photo with the birds in the background. I felt it would only work, however, if I was wearing it. In temperatures of −30, wearing a thin, short-sleeved cycling jersey wasn't one of my best ideas. One morning, I put the jersey on underneath all my thermals. It was a clear day with a bit of wind, extremely picturesque yet extremely uncomfortable. Stefan grabbed his camera. 'Are you ready?' I shouted. Stripping off, I exposed my arms. Immediately, the hairs stood on end and froze. A stinging sensation shot through them. I rubbed them up and down hard while Stefan clicked away but after no more than twenty seconds, I was done. Whether Stefan had the shot or not, I couldn't handle it; it was just too cold on my bare skin.

Luckily, he got one in which I was smiling and with the club's name clear as day on my thin jersey. Despite appearing happy, I was in agony and I just hoped it was worth it and that it was something different for everybody to appreciate when I eventually emailed it over that evening. Part of me wished I was barbecuing and spending time with friends reminiscing about our old mate. I was relieved, however, that I wasn't getting eaten alive by midges after a long bike ride!

With the spring winds having finally released their grip on Atka Bay, I now had only up to a couple of weeks left in Antarctica. Each day that the emperor chicks got through, the higher their chances were of reaching adulthood. They'd grown rapidly over the past month and were already half the height of their parents. Their target weight for fledging was around half their parents' weight, approximately ten to twelve kilograms, and they still had some time to go.

There was not much going on around the colony other than chicks feeding, which gave me the opportunity to head back over to the cluster of icebergs in better weather to film the scenery and beauty of the temporary landscape the penguins were living on before it broke up. Having passed the spring equinox, days were finally getting longer than nights and with the sun once again only dipping below the horizon, darkness was now a thing of the past. Using every available minute of sunlight, I worked harder than I'd ever worked before. Splitting my time between my bed, my skidoo and the ice was physically exhausting, but worth it.

Stefan and I had spent a long day capturing the sun rotating round the towers of ice that angled out of the 'Yosemite' iceberg. The sun set at 10pm and we'd made our way back to the station, about an hour's skidoo drive away. Both exhausted, we grabbed some tea and said goodnight. But with so little time left in Antarctica and an exceptional much-needed run of good weather, I was keen to get back out to the icebergs. They were just too inviting. The sun was due to rise again at 2am and I couldn't bring myself to wake Stefan after such a long day. Convincing Will it was worth it, I grabbed an hour's sleep, refilled my flask and left a note outside Stefan's door.

The sky was pink and the air was fresh as Will and I drove our way back along the tracks I'd laid just three hours before. Will had only managed a couple of trips over to the icebergs during the winter and seeing them under a different light made the landscape look surprisingly different. Even though I'd spent the day amongst them, I was desperate to get back underneath their vertical walls before the sun reappeared. As we sped side by side across the ice, I looked across to Will. The plume of snow behind him made it look like he was burning the ice as he raced across its pristine, sparkling smoothness. The sky in the north was a beautiful peachy hue that melted into the deep blue above our heads. To the south-east, the sky's fiery orange glow signified the sun wasn't far below us.

We parked our skidoos and raced to get the camera gear ready. It felt pleasant but with the sun still below the horizon, temperatures had fallen back down to around −25 overnight. As soon as I was set up, looking down a narrow avenue between two tabular icebergs, the sun began to reappear. All of a sudden, the flat snow beneath my feet became a carpet of glitter as each individual crystal of ice reflected the sun at a different angle. Never-ending sweeping shadows began to shorten. With the penguins away in the distance and not a breath of wind, for the first time the surroundings were silent. An indescribable silence. My ears started to ring as if to confirm to themselves they were still functional. I'd never heard silence quite like it.

I closed my eyes and tried to envisage what life was like right at that moment back at home. I tried to feel the sensations of normal life, sitting in a car in traffic or by the river near my house, but I simply couldn't. I'd forgotten. I turned to Will; we were both in a world of our own, struggling to believe we weren't dreaming. Where were we? Had we really just spent a year living with the emperors?

I'd had very few opportunities filming alone with Will during our time in Antarctica, and being the only two Englishmen and having got to know each other so well during our time down there, watching the sunrise in those circumstances felt special.

The landscape we were sitting in felt as though it belonged to us. It hadn't been touched by any human before and would never be touched again. Within a month or two, it wouldn't even exist and we realised and appreciated just how incredibly lucky we were.

By 6am the sun was high above us and we were both drained. I'd become used to ignoring the difficult conditions we were working in, as well as the long hours. Even though spring had returned and long days were beginning to warm the atmosphere, it was still challenging, despite my body having become accustomed to it. Having to breathe through my balaclava, operating delicate bits of kit with my enormous mittens and having my body weighed down by all my thermal and polar clothing was the new normal. Driving back towards the station to grab some breakfast before going out *again*, I started to feel the strain on my body. Even at −25 degrees, bouncing over areas of lumpy sea ice on the skidoo, I was falling asleep. I was pushing myself hard but with only a few weeks left in Antarctica and a remarkable run of good weather, I just had to make the most of every second of each filming opportunity. Although I didn't have long left, the good weather meant I really didn't have any time to think about the prospect of finally going home.

8.

My Final Few Weeks

O ctober was coming to an end and so was my Antarctic adventure. The weather had been consistently good for three weeks, giving me the opportunity to split my time between the icebergs and the penguin colony, and I'd tirelessly made the most of it. Not only was I desperate to maximise my time for filming, I was also acutely aware there was a good chance I'd never return to this place so was keen to take in as much of my experience as possible. The scientists and station staff that I'd spent so long with still had over three months to go before leaving Neumayer. For them, Antarctic life continued as normal, but they were preparing for station life to change. After eight months alone, they would soon have to start sharing the station that we now referred to as our home. In the blink of an eye, the frenzy of a summer season and up to sixty other personnel would be on top of them, just as we'd arrived into the previous team's world.

Our routine and way of life was consistent and we'd become used to living in each other's pockets. We'd helped each other where we could, supported and laughed with one another, and despite the worrying prospect of altercation and arguing, we'd got on remarkably well. We'd managed to get through with only small disagreements and there hadn't been one argument or falling-out that I was aware of.

I was really proud of this. Prior to our experience it had been one of my main concerns. I'd read and heard some horror stories from other overwintering stations where personnel had barricaded themselves into their rooms following arguments, on some occasions due to violence. Only recently, a stabbing had occurred at a Russian station. Isolation did weird things to people but thankfully we all trusted each other and had become an amazing team. Dr Tim's leadership had been exemplary and somehow, after a near fire, numerous late-night parties and several unicycle collisions, we'd kept the station in the same condition in which it had been handed to us. In fact, with a couple of last-minute deep-cleans that Tim had ordered, with the help of the whole team, Neumayer was in fine fettle.

During the dark days of winter, when my morale had been at its lowest, our team back in the UK had begun making plans for our return journey. Will had been toying with the idea of a few weeks' holiday in South Africa en route to home. Stefan was keen to head back to Germany and I was obviously desperate to return to the UK to my new family. Somehow we also had to get our fifty-plus camera cases back and the all-important emperor footage we had stored on numerous computer hard drives. It was as logistically difficult as organising our journey south had been, and needed thorough planning.

The team had been given almost two years in which to make arrangements for the journey that had started our Antarctic adventure, yet they had only a few months to work out how to bring us home again. They were highly experienced, however, and I had every faith. I'd always been promised the first flight out of Neumayer at the end of the winter isolation, and during the dark days I felt that day couldn't come quickly enough. Without sunlight, those dark days had felt endless. With my sleeping problems and spending so much time outside in the freezing conditions, my body had felt the full effects. I was absolutely exhausted.

Even though I'd managed to keep my mind busy, at times I just wanted the whole thing over with and to get home on that first flight. Being back with Becky in the warmer fresh air of the Lake District was all I craved. Even when I had managed to get out to film over the winter, progress had been slow; the male penguins had not been doing a great deal and the cold conditions had been horrendous.

Although we were unsure when a plane would actually reach Neumayer station, dates and flights between Novo, the Russian airbase, and Cape Town were already scheduled. Dr Tim had printed off a flight plan for us to look over. The first flight, to my shock, appeared to be in early December, well over a month after what I'd been promised. I couldn't understand it. The international flights had always operated from as early as possible in the season, sometimes as early as late October, so I asked Tim to double-check. I didn't dare tell Becky as she'd had early November in her head from the day I'd left. In the grand scheme of things an extra few weeks on top of eleven months didn't appear much, but having that date cemented in my mind, I didn't think I could stay any longer. Eleven months was my limit. The prospect of anything extra made me feel sick.

After a few days, an embarrassed Tim realised he'd made a huge mistake and that he'd printed off half of a previous year's flight schedule. How, I have no idea. He'd mistakenly presumed that the top of the list was the first flight out. Discovering the error, a feeling of relief came over me. I'd already warned Becky's mum that an extra month was looking possible. She'd laughed it off; nothing surprised her any more. I hadn't been able to bring myself to tell Becky I was staying longer. I don't know what she would have done. Knowing Becky, she would probably have tried to charter a plane to come and pick me up. Luckily, now I could message her mum to say there had been a mistake and I would hopefully, depending on the weather, be home as scheduled in November. Messaging Becky's mum to say there had been a mistake was a lot easier than going through the process with Becky herself. I felt very lucky that I could chat through things with her without having to upset or worry Becky.

Spring's increasing day length brought physical and psychological benefits and finally a feeling of normality was slowly returning to my body. In fact, I was thoroughly enjoying every day. I still thought about home constantly. Having distractions made time go a lot faster and I was really appreciating the privilege of where I was and what I was doing. The sea ice of Atka Bay and the emperor colony felt like my second home and our team of twelve people living on the station felt like family. We'd sat down together around a table every day for dinner. We had drunk with each other every night, watched films, played games. We'd looked after one another.

Despite the weather having disrupted filming sometimes, everyday life had been incredibly consistent. Money didn't exist, time wasn't important and it never mattered what day it was. I had got very used to only having to think about myself and not having to worry about anyone else. With the prospect of returning to reality getting ever closer, I started getting incredibly nervous. I hadn't seen anyone other than my Neumayer family for eight long months. I hadn't seen a car, a tree or experienced the rain I was so used to back home. The thought of all this hitting me without having time to prepare was terrifying.

I'd read some cautionary tales of people returning to reality having spent time in isolation. I worried that leaving the station and getting back to the real world would affect me as it had others previously. But for the majority of the time, I hadn't felt trapped during my time in Antarctica; some of the team had been stuck in the station a lot more than me, and my opportunities to get out and about had definitely made a difference. The incredible communication facilities that had enabled me to call family whenever I wanted and the ability to keep up with global news meant I'd not felt the 15,000 kilometres away that I really was. Even though I was at the other end of the planet, I sometimes felt like I was just down the road.

Not that many years ago, before satellite internet had made it to Antarctic research stations, overwintering personnel were given a limit on the total number of words they could fax back home to family and friends. There was no internet. Isolation meant isolation. In comparison, I'd had it easy and incredibly comfortable. Thinking back to those sleepless nights a year before when I'd worried about being away from family and friends, I felt relieved I'd gone ahead with the trip.

The time difference was another huge factor that had helped me stay so connected to family. There was only an hour's time difference between me and the family at home, so it felt like we were all living the same lives, just at a distance. Becky and I could chat through the day, unlike on previous trips where I had been awake while she was asleep and vice versa. We didn't have the problem of staying up late or getting up early just so we could have a conversation. We were always connected. Becky's body clock had been turned upside down with Walter anyway, but while we were apart, it made our lives that little bit easier.

After 892 days since being asked to document the lives of emperor penguins across an entire breeding season, I was told that in less than forty-eight hours it would all be over. Due to changing weather conditions across Queen Maud Land, a passing plane on its way to Novo, which required refuelling, was due to land and take me away almost a week earlier than expected.

This was my transport for the first leg of my journey back to civilisation. The weather had been kind but with a deteriorating forecast the pilots were keen to get their aircraft over to Novo where their summer season would begin in earnest. It had taken them about ten days to fly down through North and South America from Canada and for the next four months Antarctica was where the plane would stay, transporting people between stations and remote camps.

Almost immediately the mechanics got to work re-grooming the runway, which hadn't seen any action since February. Flags that from a distance once again resembled penguins were repositioned in two parallel lines from west to east. The two mechanics, each in a snowcat, drove up and down continuously ploughing and re-flattening the airstrip in preparation for the plane's landing. I began furiously packing camera equipment, trying to get the mountain of kit back into the same boxes in which it had arrived. Some kit would fly back and some would be shipped back aboard *Polarstern* when it arrived in the new year.

To enable me to film up until the very last minute, Will had volunteered to remain at Neumayer for a few extra days until the following plane could pick him and the kit up. His plan was to finish the packing so I didn't have to eat into filming time to do so. Even so, I packed as much as I possibly could to save him work in his final few days. Stefan, who was travelling with me, loaded his own bags and boxes and suddenly everything seemed to have become urgent. I'd been there so long, the process of actually leaving hadn't crossed my mind. I started to get excited.

During an unsettled day at the station I made good progress and lined the fifty-plus camera boxes up along the corridor outside the kit room. I disassembled my unicycle, packed away my fly-tying kit and wrapped my three golf clubs in bubble wrap. There weren't many books I hadn't touched throughout the year on my bookshelf and as I boxed them up, I flicked through their pages. A book on British birds, moths and dragonflies, and some cyclists' autobiographies. The last book on the shelf was a BBC book entitled *Planet Earth, The Making of an Epic Series*. On its front cover was an image of the wildlife cameraman who had filmed the sequences in the programme, on his knees in front of enormous icebergs surrounded by emperor penguins. It was the same scene that had sucked me in all those years ago. I'd not read the book all year, forgetting it was there hidden behind some German language guides.

As I sat on my bed and flicked through to get to the Antarctica pages, a note fell onto the floor. Dated 1 July 2006, it was from my old physics teacher. It read:

> Lindsay,
>   Hope you haven't already got this! Just wanted to wish you all the best in the future. Maybe one day your work will be in a book like this!
>   Keep in touch,
>   Mr Rogers

I had been just sixteen at the time, about to sit my GCSE exams. Little did I know that the next time I read the note, I'd be in the place I had dreamt of reaching.

My final day arrived and with it one of the most pleasant days since landing the previous summer. Having not seen the penguins for twenty-four hours I was keen to say goodbye properly and wish the chicks luck for the remainder of their juvenile life. With no sledge piled high with camera kit and no rucksack on my back, the skidoo felt light and in under ten minutes I was sitting in front of the colony. As usual, chicks and adults wandered over for company. The chicks were looking fat and strong and on such a calm day it was hard to recall the brutal conditions they'd already had to endure to survive to this stage.

Hovering over the colony, a couple of Antarctic skuas had returned and were on the lookout for any spilled squid or fish from the adults. Even though the weather was pleasant, they still had to be quick, picking their moment to glide down into the colony and grab the dropped food off the ice before it solidified and became inedible.

In the middle of the colony a couple of Adélie penguins had returned and started to build a nest out of lumps of ice. Just like the unstoppable maternal instinct the emperors had shown with their snowball-egg, the nest of ice was something that would prove useless, but in a similar way they were practising for the real thing. I took a walk around the birds and up onto a large snowdrift to look over the entire colony. A couple of adults followed me up. The dreaded gully had thankfully been filled in and all the bodies of perished chicks and the scattered unhatched eggs lay hidden deep under layers of snow. I looked back over towards the ramp that led up onto the ice shelf to remind myself that I was standing on top of an ocean. Because I couldn't see the deep blue water beneath my feet, I'd gone the whole year forgetting where I actually was. Even the station base, floating on an ice shelf above hundreds of metres of water, was hard to comprehend.

Within a couple of months, the sea ice in the bay would break up and the penguins would have gone. In their place there would be orcas and minke whales, Weddell and leopard seals and waves. Having spent so long with the penguins and gone through so much together, I didn't feel like this was the last time I'd see them. Before I'd travelled to Antarctica, a lot of people had asked me if I was worried about getting bored of seeing only emperor penguins for such a long time. It had actually been something I'd worried about myself but in the end I'd found it far from boring. Every day that I'd spent with them had been different and I'd learnt so much. I felt like an expert on their behaviour. Driving back up onto the ice shelf to head back to the station for lunch, I stopped and took one last glance back at the colony, just in case I never got the chance again.

With only a couple of days to think about leaving, it had all been a mad rush to get ready. I knew I'd find departing difficult but in the end, having so little time to think about it helped with my emotions. At 7pm that evening out of the clouds appeared the same plane that had brought me to Neumayer. Her flanks were striped in blue, red and white and her name, *Lydia*, was printed on her nose under the cockpit window. It felt like yesterday that she'd dropped me off to begin my Antarctic adventure and it seemed appropriate she'd end it too.

Strolling down to the plane, I stopped off next to the satellite radome where the wireless internet signal was just as strong as it was inside the station. I video-called Becky to let her know I was on my way. She was getting more and more excited that I was actually coming home. I spun around to show her the landscape, the plane and Neumayer one last time. There was a decent chance I wouldn't be able to talk to her once I reached Novo and with the possibility of being there for a few days, I didn't want her to panic. I'd been told the pilots had been given an option to overnight at Neumayer rather than completing their journey to Novo in one day, and part of me wanted them to stay, giving me those extra few hours at the station. But they were keen to finish refuelling and it was clear they wanted to get the trip done in one go.

A huddle had formed around the steps of the plane. The entire team, apart from our IT technician who was manning the radio inside, stood ready to say goodbye. As when I'd said goodbye to Becky, I hugged each one individually as hard as I could. Dr Tim placed his hands on either side of my head and told me to enjoy every moment with Walter. The whole team had grown almost as fond of him as I had, and had asked for updates and photographs almost every day since the day he was born. They all wanted me to send them pictures of me with him as soon as I returned.

Lastly, I grabbed hold of Will. Having started the whole process with him almost three years before, I was upset not to be finishing it alongside him. I completely understood, however, his professional approach and his wish to get the very most out of the penguins, and through my teary eyes I wished him a safe flight out and told him I'd see him back home.

Sitting next to the window as we taxied to the end of the runway, I pressed my forehead to the glass and took one last look back at my team and the station. With the snowcats' hazard lights flashing and everyone waving madly, we took off into the air. As one final treat, looking north out of the window, Atka Bay opened up below me. The dark patches of the emperor penguin colony looked just inches from the station and the enormous icebergs I'd had such memorable moments underneath appeared as specks across the huge expanse of sea ice. To the north, further even than I'd imagined, was the open water. I took comfort in knowing it was just too far. I'm glad we tried, though.

At approaching midnight, we landed at Novo. The sun had only just set but the moon was bright against the lavender sky on the horizon. I'd completely forgotten what Novo looked like. Having already had one visit from the Ilyushin cargo plane from Cape Town, the ice runway was pristine and the ground staff were well rehearsed for their new season. Lining the side of the airbase parallel to the runway were ten containers, each with four bunk beds inside, providing temporary accommodation for people held up by weather awaiting further travel. The Ilyushin wasn't due to return from Cape Town for another two days so Novo was my interim base while we waited. Stefan and I grabbed some dinner and went to bed. It had been quite a full-on day.

Despite fresh food having arrived at Novo with the first flight from South Africa, the Russians had a unique way of cooking it. The months of tinned and frozen food had become mundane and I was desperate for a fresh salad and some vegetables. It quickly became apparent that Novo wasn't going to provide me with that or a decent cup of tea and the time waiting at Novo seemed to take forever.

It was strange to be around the airbase as it was so completely different to Neumayer. The skidoos looked like something out of the 1980s, the heavy-lifting machinery used to unload cargo was old, rusty and well used and the lavatory was a compost toilet, not much use in temperatures hovering in the minus mid-teens. There were also no showers, which, after three hours in a hot plane wearing polar clothing, didn't feel great. Our sleeping container was huge and with just one steel electric radiator, it was freezing. I tripled up my blanket by taking the ones from the bunks that weren't in use, but it still wasn't enough. All I wanted was to get home.

As well as *Lydia*, which had brought Stefan and me to Novo, a smaller plane had made the same journey down from Canada via Neumayer. Although it was due to end up at Novo to begin its summer season, being smaller and not able to fly at the same speed, they'd landed late at Neumayer and decided to stay the night and fly the following morning. All had gone as scheduled. Now, arriving at Novo, the pilot came straight up to me with a large brown envelope. Despite me triple-checking I had everything before leaving Neumayer, Will had forgotten to give me the carnet, a document without which I would be unable to get the camera gear and film footage back into the UK! It was an enormous stroke of luck there was a second plane to get the documents to me. The last thing I wanted was complications during my long journey home.

Mid-morning on the first day of November, Ilyushin appeared above the horizon in the light-blue sky. There wasn't a cloud in sight and the sun's strength once again proved too much without my sunglasses. After it landed on the strip of ice, I watched as the passengers climbed down the steps onto the snow. It was obvious that, for most of them, this was their first time. I tried to remember what I had felt like when that was me. Although it was almost a year before and so much had happened in between, it felt like only a week ago.

Just as when I had arrived, Novo's staff were quick and efficient to unload the plane's cargo and manage the scattering of people taking photographs that gathered around the nose of the plane. I wondered if I'd recognise anyone as I'd been informed some of the people who had left Neumayer back at the beginning of the year were returning to work for another summer season. But recognising people who were wearing hats and dark sunglasses was impossible.

Throughout the afternoon as our bags were loaded into the plane, I strode up and down outside our container absorbing the last bit of Antarctic sun before leaving. Over dinner I was handed the most informal boarding card I'd ever seen, a business-card-sized piece of paper with a scribble on it, and told to be on the plane by 11pm for an 11.30pm departure. The pilots had been resting since landing and the seven-hour flight overnight would get me into Cape Town early in the morning.

Late the same evening I took one last stroll around the aircraft and climbed aboard. The plane was empty; Stefan and I were its only passengers. The flags that had lined its inner walls had all been removed and there wasn't a single seat. There was also no cargo, but the Portaloo remained in position ratchet-strapped down, albeit with a new paint job. Ironically it had been shrink-wrapped in an image of an emperor colony with chicks the same age as the ones I'd just left. Our bags were in a small pile in the middle of the plane, laid on top of an emergency ladder to stop them sliding about. I selected a spot on the floor along the left-hand side and sat down. Feeling shattered, I knew I'd need to get some sleep before landing in Cape Town. To my surprise, the spare wheel, which also remained ratchet-strapped to the side, doubled up as a reasonable pillow, albeit a rather hard one. As I breathed my last mouthful of Antarctic air, the door was closed and within a few minutes the plane was airborne.

Incredibly, I managed to sleep intermittently throughout the flight and landing in a dull, damp, humid Cape Town didn't come as quite the shock I thought it would. However, the air immediately felt dirty as I climbed down the steps onto the tarmac. The air I'd been breathing for the last year had been the purest that planet Earth had to offer and my taste buds could detect the difference. I'd been told to expect possible difficulties re-entering South Africa and the UK having been away for almost a year, and I had special letters ready to show the customs officers, but luckily my passport was checked and I was thrown back into the real world without any issues.

Although I'd seen a few other people at Novo, Cape Town International Airport was my first real encounter with reality and other human beings and I wasn't sure what to expect. Busy places had always slightly overwhelmed me, but in the end, my rapid transition from Antarctic isolation into the real world was thankfully quite easy. Being early morning, the airport was relatively quiet.

The previous week it had been arranged that I'd stay at a hotel overnight in Cape Town, ensuring if my flight from Novo was delayed I'd not miss my flight back to the UK. But as everything had gone to schedule I managed to bring my booking forward a day and get a seat on the evening flight back to London the same day. It meant twelve hours in the airport but I was happy with that if I was able to jump on an earlier plane. I said goodbye to Stefan, who had taken the offer of the hotel. He also was desperate to get back to his wife in Germany, but had the agonising wait of an extra day to cope with. I waved him goodbye and went to find a cup of tea with some fresh milk.

Becky had been tracking my flight out of Antarctica online all night and, having not been able to speak since I was at Neumayer, we were keen to connect. Sitting down with two huge fresh fruit salads for breakfast, I attached my phone to the free internet in the airport. It went ballistic. I received messages from friends, family and colleagues wishing me luck with my return journey, all keen to hear about my reunion with Becky and introduction to Walter.

I hadn't stopped thinking about Becky since the day I'd left her and even eleven months down the line, the last image in my head of her standing at the front door waving goodbye was as vivid as when it had happened. I had no idea how I was going to react to her or Walter, and the anticipation was intense. To appear as attractive as when I'd last seen her, I'd made more effort than normal. Ten days before, at Neumayer, I'd cut my hair to the best of my ability to give it a few days to recover and look more natural. I'd clean-shaven rather than clipped for the first time in a year but having not showered since leaving Neumayer, I didn't smell great. Luckily, attached to a public toilet in the terminal was a cubicle with a free shower. Searching the aisles of the duty-free gift shops, I found a small block of soap, but the nearest I could find to a towel were two square flannels, one red, one blue, both with the South African flag stitched across the centre, designed for wiping mud off the end of golf clubs. It was worth it.

After over an agonising twelve-hour wait, my flight began boarding. I sent Becky one last message saying that I'd see her in the morning. She was ecstatic and told me she had 'Walter's meeting Daddy outfit' laid out on the bed ready. What that meant, I didn't know, but I was becoming just as nervous as I was excited.

Squeezing my way down the full plane, I reached my seat, a bulkhead aisle seat with two occupied seats to my left. I noticed a baby on the fold-down platform in front of the occupied seats. I couldn't believe it. The one flight when I was in need of some sleep and I'd been lumbered with what appeared to be a newborn baby. I briefly considered asking to be moved, but as I looked at the sleeping child I had a reality check. I had no idea how old he was or any concept of the difficulties of parenting, let alone travelling with an infant. In just a few hours I'd be in a similar position, just not on a packed plane. Embarrassed with myself for even contemplating moving, I settled down and started to talk to the couple next to me. Ironically, they were German and their baby was seven months old, almost exactly the same as Walter. I explained my situation and where I'd been. Even though I'd seen hundreds of pictures and videos, I had no sense of how big Walter would be, so seeing their little boy fast asleep gave me an idea. They were a lovely young couple and throughout the flight I interrogated them about the first few months of parenthood. Their little boy was beautiful and didn't make a single sound throughout the journey, leading me into a false sense of security!

When we landed on the Heathrow runway, I hadn't slept at all. My stomach had churned all night on the flight with the nervous anticipation of being reunited with Becky. I'd turned my phone on but was keen not to talk to her until I actually saw her face to face. I wanted it to be special. She texted to say she was waiting for me and that she could see on the arrivals board that my plane had landed.

Racing through to the baggage collection area, I waited for what seemed like an age for my six cases to arrive. In addition to those was a piece of hand luggage I had not let go of: a small black plastic case that contained the film footage I'd spent all year obtaining. I'd brought half of it back and Will was due to bring the other half. As a safety measure we'd backed it all up onto another set of drives, which were coming back on the ship with some of the camera kit, but I still felt an enormous pressure not to damage or lose it. I wanted it out of my hands as much as I wanted Walter in my hands.

With all my bags finally piled up on a single trolley, I took myself off to one end of the baggage hall where it was quiet to record a little video to give to Walter when he was older. Talking into the camera on my phone as if I were talking to him, I explained how incredibly grateful I was that he had been such a wonderful baby for his mother. In my head, he'd spent the first seven months of his life without a dad, but I was about to change that and I wanted him to know what that meant to me. He was too young to remember anything that was going on but I was sure that knowing where his dad had been when he was born was going to be a big part of his life as he grew up. As I spoke to the camera, emotion took over and tears started to stream down my cheeks. The enormity of the whole occasion started to become apparent.

I had one last thing to do before meeting him. Reaching into my backpack, I grabbed Walter's penguin, which hadn't left my side since he was born, and sat it on the top of my luggage. After I'd raced through customs and got the documents signed that Will had managed to get over to me at Novo, I was free. More than an hour after landing, I wiped my eyes one last time and pushed my trolley through the arrivals door.

There they were, running towards me, Becky and in her arms, little Walter. I let go of my trolley, still in motion, and wrapped my arms around her, squeezing as tight as I had the day we parted. 'I'll never do that to you again,' I mumbled into her ear. Releasing my grip, I grabbed hold of Walter. He was dressed in a shirt patterned with swimming Atlantic salmon with a grey tweed waistcoat over the top. Even though he was seven months old, he was the youngest baby I'd ever held. He felt solid, a lot smaller than expected but thankfully unafraid of the stranger whose arms he was in. Despite never having touched Walter, I felt like I knew him as well as Becky did. I gave him his penguin and tightened my grasp around his little legs. The sick, nervous feeling deep in my stomach released itself and the reunion with Becky made me feel as if I'd never been away. It felt natural. She looked fantastic and healthy, which is more than could be said for me. With the physicality and extremity of my time in the south, I'd lost a stone.

'You don't look well,' Becky said to me with her usual sarcastic humour. Nothing had changed.

Postscript – Back to Reality

Only thirty-six hours had passed since standing on the ice at Novo looking over the Antarctic landscape, but already it felt a million miles away. Just as when I'd sat underneath the 'Yosemite' iceberg struggling to envisage what life was like that very moment back home, I already found it impossible to remember what it felt like standing on the ice in −30 temperatures in complete silence. Bad weather had hit Neumayer as predicted, and Will was trapped for the time being.

The transition from my independent life in Antarctica back to reality had been quick and, surprisingly, I didn't experience the difficulties I'd expected. Meeting and talking to different people every day on the street, dealing with traffic on the roads, being responsible for money, having jobs to do, the sounds of civilisation, all just felt very normal. I slipped back into life without any issues. Walter kept me busy; having him to concentrate on distracted me from any difficulties. There wasn't time to sit around thinking about Antarctica.

Probably my biggest surprise was returning to the British weather. I'd spent eleven months in temperatures that had only just risen above freezing three times. The average temperature across the whole trip was around –20 degrees Celsius. My body had adapted and become so used to such severe temperatures that I could walk around outside in Antarctica in temperatures down to –15 in a normal set of clothes. At home in the Lakes, at that temperature I would have been in full thermals.

So I'd expected that, back home, I'd be wearing shorts and a T-shirt even on frosty December mornings, but to my surprise I'd gone the other way. Every part of my body felt extremely sensitive. Antarctica's air was dry so, although it was freezing cold, it didn't get through to your bones like the damp air does in the UK. British rain was something to which my body needed to reacclimatise.

After I had been home for a week, Will was still stuck at Neumayer. I'd been talking to him every day and even though my time in the south had come to an end, I couldn't quite let go of old habits. A couple of times every day I'd check the weather forecast online, even downloading any up-to-date satellite images that had appeared to see if the condition of the sea ice had changed. It was obvious that the weather system that was stuck over Neumayer wasn't easing up. However, after almost ten days of contact, Will suddenly went quiet. I could see the weather had calmed down and I wondered if, at short notice, a plane had arrived to fly him out. I sent a couple more messages. No reply. If Will had managed to leave it would be at least twenty-four hours before I heard from him again, as there was no way of contacting me through Novo until he reached Cape Town.

That evening my phone vibrated with a message from him. 'I did it. One of the toughest things I've ever done,' he wrote, '26 miles. 5 hours, 36 minutes.' While I had been drinking tea and watching telly, Will had completed his Antarctica marathon. Grabbing the last and only opportunity he had with a break in the weather before leaving, Will had filled some flasks, devised a route across the ice and set off. I was so proud of him. It was such an amazing thing to do and the achievement was even more impressive given that we were at the end of such a long and arduous trip and seriously lacking in energy.

Within two days, Will was on a plane back to London from Cape Town. The calm conditions had lingered, allowing a plane to scoop him up from Neumayer along with all our kit, to get him to Novo to catch the international flight to South Africa. Unfortunately, as he'd been stranded for the extra two weeks, the weather had eaten up all of Will's planned holiday time.

Three weeks after returning, I began to feel better both mentally and physically; the exhaustion was wearing off. Starting to feel more like myself, I was lulled into a false sense of security. Unbeknown to me, my body's immune system was weaker than it had ever been. With the last of the summer staff having left Neumayer back in February, nobody had entered our sterile world since then, so our immune systems had not had to fight any new bacteria. Slowly, my resistance to germs and viruses had become weak. Gorging on food that my body wasn't used to and being exposed to other people and germs hit me hard. What started with vomiting for forty-eight hours was followed by flu and then every kind of cold imaginable. I'd been warned I would experience some effects on returning but it was far worse than I thought. It wasn't until after four long months that I finally felt like I was recovering.

In late January, after a first Christmas and New Year as a family at home, I spoke to Dr Tim on the phone. I'd been back a couple of months but Tim still had a month left at Neumayer. He'd been keeping me updated on what was happening at the station, who had returned to work the summer season and any trips he had made to see the penguins.

On 27 January, however, he sent through an image I hadn't been expecting. It showed four flags marking out the top of our ramp, the edge of the ice shelf and an ocean full of open water. Atka Bay had broken up. The seemingly never-ending landscape I'd spent so many magical moments on was no more. No icebergs. No penguins. I had spent every day on the surface of a frozen ocean, but the fact I couldn't see the water beneath my feet had made me unaware of the fact. The breaking up of the ice emphasised the landscape's fragility and made it seem miraculous that I'd spent so long above hundreds of metres of water.

I just hoped the emperor chicks had managed to moult into their adult feathers before the ice broke up. If they had been forced to fledge with their grey downy coats, they wouldn't have stood a chance. It was the second year in a row that the ice covering the bay had broken up, following years of stability throughout the summer. Whether this had an impact on the success of Atka Bay's breeding season was hard to tell, but it certainly didn't do it any favours. According to an expert in the United States who'd studied emperors his whole life, the time the bay had remained frozen should had been long enough for around a third of the chicks to have managed to fledge. But what about their future? Not all of the colonies around Antarctica had been so lucky.

In Halley Bay, not far from the British Antarctic Survey station, the emperors have had a tough time due to changing sea ice conditions. For several decades, the population breeding alongside the Brunt Ice Shelf had numbered an average of 14,000 to 25,000 breeding pairs (5–9% of the global population). With sea ice breaking up before the emperor chicks had developed their adult feathers, which would have enabled them to swim by protecting them from the cold water temperatures, they'd all perished. After a couple of failed breeding years, the adult birds had realised the location was no longer reliable for breeding. The colony had to make a choice to either not breed or move. The Halley Bay colony, the world's second largest, effectively collapsed.

At the same time, the Dawson-Lambton Glacier colony fifty kilometres to the south-west saw a big rise in numbers, suggesting that many of the emperors had relocated, although not all of them. Whether the change in sea ice at Halley Bay is solely down to climate change isn't completely understood but if the trend in the loss of sea ice continues, computer models envisage emperors might lose anywhere between 50 and 70 per cent of their global population within eighty years, a prospect I can't bear to think about. I want to take my children to Antarctica one day to show them the amazing emperors that mean so much to me. The thought that they may not be there brings tears to my eyes.

Before my time in Antarctica, I'd found it hard to understand why so many people chose to return to work there year on year. The purpose of my visit had been completely different to most jobs around a research station. I couldn't really compare myself to those people, but having spent a year at Neumayer, I totally got it. The place had me hooked. Out of the 337 days I spent in Antarctica, we were isolated for 245 of those days, and over 100 of those were spent inside due to the weather. Everything was as extreme as I will ever experience but I wouldn't have changed a thing. As for the emperors, having spent so long with them and witnessed almost their entire life cycle, I cannot imagine living the rest of my life without seeing another one.

When I was out there filming I never really knew what the finished film would look like. I literally handed over my material and hoped I'd managed to capture it all in enough detail for it to be edited together. It was a terrifying wait and I always dreaded a call that may say there was something wrong with the footage or that I'd missed something important. As it was, the team did an incredible job with what we'd managed to capture and the film was simply stunning. Eight million people tuned in to watch the second episode of *Dynasties* in the UK when it was first broadcast and since then, it has been shown across the world.

I still find those numbers difficult to comprehend. Very, very few people get the chance to visit Antarctica, let alone see an emperor penguin in the wild. I just hope the film demonstrated to all those people just how incredible a creature they really are. A quote I read before I believed I would ever make it to Antarctica had always struck a chord with me. Apsley

Cherry-Garrard, an English Antarctic explorer and author, wrote after his expedition in the early twentieth century, 'Take it all in all, I do not believe anyone on Earth has a worse time than an emperor penguin.' I can now say I have witnessed first-hand how true that statement is.

Acknowledgements

This book would never have been possible without my agent Annabel Merullo at Peters Fraser and Dunlop and my publisher Rupert Lancaster at Hodder & Stoughton, who not only gave me the opportunity but encouraged me to turn my extraordinary year into words.

From a filming and career point of view, one of my favourite parts of my work is meeting the incredibly passionate people who not only make my job so much easier but are helping conserve our amazing natural world with such enthusiasm and persistence. Whether I've spent five minutes with you or five weeks, I thank you.

For my first opportunities and endless advice as a teenager, I owe a huge thank you to Colin Jackson and the late Ian Dewar without whom I wouldn't have a career in wildlife filmmaking. At the BBC Natural History Unit, I must thank Miles Barton, Rupert Barrington and Mike Gunton for offering me my dream job, sending me to an emperor colony and trusting me with such a huge responsibility. The BBC *Dynasties* team that helped, from travel and arranging camera equipment, to editing the final programme, Lisa Sibbald, Alison Brown-Humes, Theo Webb, Kirsty Emery, Gordon Leicester and Angela Maddick.

At Neumayer I couldn't have been more proud of how our

team, the 37th Overwintering Crew, supported each other and made the whole experience, especially the long and mentally challenging period of isolation, so much easier than I envisaged: Tim Heitland, Zsófia Jurányi, Maximilian Merl, Ursula Schlager, Hauke Schulz, Sven Krüger, Daniel Noll, Hannes Laubach and Ronny Lebrenz. From the Alfred Wegener Institute back in Germany who assisted our film crew from all the pre-trip training to arranging our flight home: Eberhard Kohlberg, Felix Riess, Sina Loschke and Christine Wesche made logistically challenging planning seem so easy.

I'd like to give a heartfelt thanks to both Stefan Christmann, my field assistant, who spent every minute of our filming year on the ice with me, and Will Lawson, my Director, for both keeping me safe while outside and making our year with the emperors so, so memorable.

Finally, I'd like to give a special thank you to my family. My mum and dad for giving me such freedom as I grew up that helped forge a fascination and obsession with our natural world. My wife Becky who supports me in everything I do, and my two boys, Walter, who thankfully is none the wiser about the fact that I missed the first seven months of his life, and Ernest, who arrived part way through writing this book. I will continue to do what I can to protect, conserve and ensure you have the wildlife to enjoy in future years that I did as a child.

An invitation from the publisher

Join us at www.hodder.co.uk, or follow us
on Twitter @hodderbooks to be a part of
our community of people who love the very
best in books and reading.

Whether you want to discover more about a book
or an author, watch trailers and interviews, have the
chance to win early limited editions, or simply browse
our expert readers' selection of the very best books,
we think you'll find what you're looking for.

And if you don't, that's the place to tell us what's missing.

We love what we do, and we'd love you to be a part of it.

www.hodder.co.uk

@hodderbooks

HodderBooks

HodderBooks